W9-ADE-076

Homer's
Secret Iliad

Homer's Secret Iliad

The Epic of the Night Skies Decoded

FLORENCE &
KENNETH WOOD

JOHN MURRAY
Albemarle Street, London

Extracts on page 157 from *Theogony; Works and Days; Elegies* by Hesiod and Theognis, translated by Dorothea Wender (Penguin, 1973; copyright © Dorothea Wender 1973), and on page 240 from Apollonius of Rhodes, *The Voyage of Argo: The Argonautica*, translated by E. V. Rieu (Penguin, 1959; copyright © E. V. Rieu 1959), are reproduced by permission of Penguin Books Ltd.

First published in 1999
by John Murray (Publishers) Ltd,
50 Albemarle Street, London W1X 4BD

A catalogue record for this book is available from the British Library

ISBN 0-7195-5780 1

Typeset in Adobe Palatino by Servis Filmsetting Ltd, Manchester
Printed and bound in Great Britain by The University Press, Cambridge

To the memory of
Edna Florence Leigh
(1916–1991)

Ad Astra per Aspera

Kansas State Motto

Contents

Acknowledgements

The material on Homeric astronomy in the following pages is derived from the pioneering studies of Edna Leigh, although the authors take all responsibility for some expansion of her ideas.

Without the generous encouragement of Dr Betty Sue Flowers, Dr Paul Woodruff, Dr Philip Bobbitt, Dr Peggy A. Kruger and Mary B. Rogers, of the University of Texas at Austin, it is unlikely that this book project would ever have been realized. We also acknowledge our appreciation of the interest and encouragement of Dr Hertha von Dechend, of the University of Frankfurt, herself a noted scholar of great perception in this field, and that of her colleague Dr Walter Saltzer. On classical matters, much has been learned from Dr Kathleen Hull of the University of Manchester and from Ian Howarth of the Open University. Linda Simonian's advice on astronomy was invaluable, as was the artwork of Gill Garfield-Ralph.

Particular thanks are due to our daughter, Carol, and our son, John, for their cheerful support and assistance, and to Marjorie Rigby, the younger daughter of Edna Leigh, who provided translations from Greek texts.

We value the skills and assistance so willingly given by Lynn and Richard May, Caroline Davidson, John Keegan, Andrew Dalby and, at the Liverpool Museum, Martin Suggett. Thanks also to the many friends on both sides of the Atlantic who patiently listened to presentations during our prolonged attempts to find an accessible pathway into Homeric astronomy.

Finally, we are indebted to Grant McIntyre, Caroline Knox and their colleagues at John Murray, for their support and enthusiasm, and to Roger Hudson and Bob Davenport, two outstanding editors and masters of their craft.

A Journey of Discovery

Nearly three millennia have passed since Homer sat under the clear skies of ancient Greece and wove the threads of time-honoured myths and old stories into the *Iliad*, an account of a few weeks in a brutal ten-year war between Greeks and Trojans. This heroic poem is universally recognized as the first great epic of world literature. The aim of this book is to show that it is also the world's oldest substantial astronomical text, whose learning has been lost for two millennia or more. Embedded in its accounts of gods and mortals is an abundance of information about the planets, the constellations and hundreds of stars, from the brightest in the sky to those just visible to the naked eye. Homer's stories were the vehicle for preserving this knowledge and passing it down through the ages.

Very little has been known until now of the astronomical knowledge of the peoples who lived in Greece, Asia Minor, Crete and the islands of the Aegean and Ionian seas for a thousand years or more before Homer (*c*. 745–700 BC). There is, however, written and archaeological evidence of applied astronomy from contemporary Mesopotamia and Egypt, and it is certain that learning about the heavens would have been as essential for the Greeks as it was for other ancient societies. Astronomy in those far-off days was not the esoteric science it is today, and ancient peoples used their observations of the stars and the moving heavens for such practical purposes as clock and calendar, as well as in their creation myths and religions.

It is not clear when the Greeks came to possess a system of writing in which poetry could be written down, but possibly it was about the time of or a little before Homer's day. Before that, poems and stories were transmitted orally, and thus the part played by memory was enormously important. And what applied to poetry applied to astronomical knowledge too. This book seeks to show that the *Iliad* was created to preserve ancient knowledge of the heavens and is not only a poem about the Siege of Troy, but also a comprehensive record of the ancients' knowledge of the skies. It is a memory aid of great sophistication, using unforgettable narrative to fix astronomical data in the mind. The poet-singers or bards who learned stories by heart and passed them down through the pre-Homeric centuries were not just entertainers but the conservators of an extensive astronomical culture. The conjunction of Homer's poetic genius with the invention of sophisticated writing later allowed both epic and astronomy to be preserved in a more permanent fashion.

The rediscovery of the *Iliad* as a repository of astronomical knowledge had an unlikely genesis, beginning in Kansas, where Edna Johnston was born in 1916. She was one of six children raised on a farm in the south-east corner of the state, where her tenacious character was moulded by the hardships which struck her family when severe drought in the 1930s turned parts of the Great Plains into the Dust Bowl. Edna later recalled days 'when the air was so dusty that the Sun at noon looked like a full Moon in a yellow sky, days when I herded cattle across those dusty acres to whatever well happened to have accumulated a little water'. Unlike many others, however, the family farm survived, and to this day her younger brother still works the land.

Edna's parents had been schoolteachers before becoming farmers, and they fostered her passion for learning. A love of

Kansas and farming remained deep within her and later added to her understanding of rural life in ancient Greece. On a number of occasions she used her knowledge of animal husbandry and hard-learned lessons of soil erosion to make significant points relating to Homeric epic.

The small town of McCune in Kansas, where Edna grew up, lies almost on the same latitude as Rhodes, and, as above that island in the Aegean, the skies over Kansas are often clear, showing the undiminished glory of the Moon, the planets and a host of stars. This nightly spectacle was an important influence on Edna, and there was increased stimulation to learn about the skies when Clyde Tombaugh, a young astronomer and the son of a Kansas farmer, discovered the planet Pluto in 1930 – a major event in a state where the weather and crop yields were the more common topics of conversation. Edna's daughter Florence, co-author of this book, feels this discovery may have been the catalyst for her mother's interest in astronomy when she was in her early teenage years. Tombaugh later said that his long and diligent search for Pluto was better than pitching hay on his father's farm – a sentiment with which Edna would have sympathized. Her interest in astronomy continued to develop, and many years later when she visited Greece for the first time she wrote:

> Our house in Kansas had a tower with nine windows. In the heat of summer my sisters and I used to lift the mattresses from our beds and lay them there on the floor. We brought the pillows and two sheets each and opened all the windows. The blinds we rolled to the top. Then we lay down to sleep covered with a single sheet, the wind and the moonlight. We lay there with our feet exposed as well as our heads, arms and shoulders and watched the stars. The wind stirred the leaves of the cottonwood tree below in rain-like whispers. Far off a dog barked. Nearer at hand the sheep moved and we listened to the briefly agitated bulls. A mule brayed. A cricket chirped. A night bird called. We slept.

At Delphi, eight thousand miles and one degree north, and thirty years between, the Moon rose late; I went to a bench on the balcony to see. For an hour I sat with the wind, the Moon and the stars for company. The hills folded downward to the side. I had come home.

Edna studied for Bachelor's and Master's degrees in literature at the Kansas State Teachers' College at Pittsburgh, Kansas, majoring in the English Romantic poets; the insights of John Keats and the language and use of Greek sources by Lord Tennyson, in particular, were to have a deep influence on her approach to Homeric epic. One of her minor subjects was mythology, another valuable tool for her later work. With her family still suffering from the economic disaster of the drought years, Edna completed a four-year college course in three years and still found time to take up a paid position in the college library and work on the student newspaper. In whatever spare time she could find, she studied music, played the violin and sang in a choir.

She later taught at high schools in several states, and after her first marriage failed she joined the staff of a college in Miami, Oklahoma. There in 1943 she produced a student play, and in the audience at one performance was John Leigh, an RAF officer who was training at a nearby flying school. Their friendship flourished, and they were married in 1944. In April 1945, Edna, with Florence and baby daughter Marjorie, crossed the Atlantic in one of the last convoys of the Second World War, to settle permanently with her husband in England. Of the possessions she took with her, the most treasured were her books, including a hand-coloured copy of the *Geography of the Heavens*, by Elijah H. Burritt, published in New York in the 1830s, C. M. Gayley's *The Classic Myths in English Literature and in Art* and works of the English Romantic poets.

Edna's life changed again in 1949, when she returned to teaching at a school for the children of US airmen at Burtonwood, a large wartime airfield between Liverpool and

Manchester that had reopened during the 1948–9 Berlin Airlift. From then until her retirement in 1981 she taught in high schools for children of the US military and diplomatic staff stationed in England.

In 1965 Edna recorded that her study of the purpose of epic had begun more than twenty-five years earlier, when she noticed some striking similarities between the funeral customs described in the eighth-century Old English epic of *Beowulf* and those depicted in Homer. 'Although my observation was by no means original, I could neither find nor think of a thoroughly satisfactory reason to account for both authors burying their heroes on headlands so that their tumuli, as Homer puts it, "might be seen far out at sea by the sailors of today and future ages"' (*Odyssey* 24.81). Why two stories separated in time by a thousand years and in distance by two thousand miles should reflect similar images of navigation aids for sailors intrigued her and strengthened her intuition that there was more to be learned from literary epic than first meets the eye.

Her interest in Homer, once aroused, never dimmed, and her thoughts ranged far and wide as she struggled to find the key to unlock the hidden meaning which she felt sure was there. 'I wondered whether in his works we deal, not with epic as we think of the term, not with heroic narrative, but with some type of carefully devised textbook . . . Eventually I was able to identify the general subject matter of Homeric epic as astronomy,' she wrote in 1965. Precisely when her first discoveries were made is not known, but certainly by the late 1950s she had discovered how Homer used astronomical knowledge to guide ancient peoples in their travels the length and breadth of Greece and Asia Minor (see Chapter 9). From these beginnings evolved our entire understanding of Homeric astronomy.

Edna's trailblazing work shows that even in today's high-technology world there is room for the *grand amateur* to make an outstanding contribution, particularly in the interdisciplinary borderlands between two subjects – in this case, between literature and the history of astronomy. A modest but cultured woman of wide learning and vision, she was not confined by the boundaries sometimes imposed by traditional scholarship. Heinrich Schliemann, the German who in the 1870s excavated at Troy, and Michael Ventris, the English architect who in the 1950s decoded Linear B, an ancient Greek script that had perplexed experts for fifty years or more, were both 'amateurs' like her.

Edna became so familiar with the *Iliad* and the *Odyssey* that she could recite them from memory, just as did the poet-singers of ancient Greece. In the 1960s she had expanded on her first intuitive feeling that much still had to be learned about the *Iliad* and the *Odyssey*:

> The epics perplexed me, for they seemed unaccountably to reflect some indefinable meaning that was comparable, in a sense, with the faint and tantalizing whispers of melody played upon an unknown instrument during the mighty crescendos of a full orchestra.
>
> Because I was unable to isolate the nearly intangible undercurrent of idea that impressed me as being common to both poems, my initial efforts to identify Homer's major theme were extremely discouraging. I set up many hypotheses and satisfied myself that most of them were untenable. Through reading and rereading the epics I eventually came to feel almost as if I were gazing upon some enormous and faraway canvas. Depicted vaguely upon it seemed to be a shifting scene, yet one differing greatly from that conveyed by the ordinary sense of Homer's narratives.
>
> Although I believed this impression could be hardly more than a ridiculous notion, I found my new concept of shadowy form behind gossamer-like substance even more puzzling and provoking than my former idea of an elusive epical meaning. Nevertheless, by the time I could actually define the imagery that my reading engendered, I almost questioned my reason: the Homeric picture seemed to consist of points and lines, of angles and arcs, of cycles and circles, of order and time, of stillness and motion and of darkness and light.

Edna believed that the poet John Keats (1795–1821) might have experienced a similar reaction to her own on reading Homer. She wrote:

'Then felt I like some watcher of the skies,' declared John Keats in his poem 'On First Looking into Chapman's Homer', after reading the *Iliad* for the first time. Scholars have always supposed Keats to be stating that he is as excited about his personal discovery of Homer as an astronomer would be upon seeing a new planet. Extending our interpretation of Keats's line is simple enough: perhaps the scientific developments of Keats's day impressed him, or possibly only the heavens seemed to him high enough for a comparison with Homer. We have never considered whether Keats actually perceived a distant, half-blurred vista as he read and – for just one moment, for just long enough to write the words – glimpsed beyond a veil.

These words of his helped me clarify the mixed impressions engendered by my own readings of Homer. In my study of the *Iliad* and the *Odyssey* I conclude that these epics represent an ancient people's thoughts related to the science of astronomy and expressed in the form of elaborate narrative poetry.

As Edna's thoughts crystallized she began extracting layer upon layer of astronomical knowledge from the epics. Twice she began to write a description of her findings, but each was abandoned unfinished when, it is thought, new discoveries made her realize the extent of the still untapped riches of Homeric astronomy. One manuscript was called *Panorama of the Whole* and the other *Watcher of the Skies*, and when first read by the authors they created a true sense of wonder at the ideas presented in them. In *Panorama of the Whole*, Edna expounded upon her first tentative approaches into the world of Homeric astronomy and her firm belief that only when the epics were considered in their entirety could the whole astronomical picture be seen. *Watcher of the Skies* was of a more technical nature and explained her method of extracting learning from epic and explored astronomical themes. On her death these typescripts together with many hundreds of sheets of notes were left to her daughter Florence, who together with her husband, Kenneth, then spent seven years editing and expanding them. Kenneth now takes up the story:

I first met Edna in 1958, and for more than thirty years this strikingly attractive woman gave an unmistakable impression of warm generosity, polite but firm authority, and great intellectual presence. The second time I met Edna, she stood by the fireplace with a sheet of graph paper in her hand, and showed me sketches of how the shapes of constellations could be used as maps to guide ancient travellers in Greece. Despite my ignorance of both Homeric epic and astronomy, I was a sufficiently curious journalist to know there was a good story to be told from her work. Since Edna's death in 1991, Florence and I have searched in vain for that piece of blue-lined graph paper. At that second meeting she also made an unforgettable remark, almost as an aside to herself: 'When Homer wrote of the wine-dark seas, he wanted us to look at the heavens, not the oceans.'

In the years that followed, Edna pursued her work almost in isolation. A private person, she was reluctant to take anyone into her confidence, and her family was rarely allowed a glimpse into her enchanted world of Homer. From time to time, she told me about her progress and latest ideas, but although I listened with fascinated interest my role was only that of a sounding board for her developing thoughts.

Over the years Edna expanded her own knowledge of literature, mythology, languages and science, and in a personal profile she wrote for a career appraisal in 1968 she said she had studied 'Greek, both classical and archaic, . . . Latin, comparative Indo-European grammar . . . some Sanskrit' and had an 'acquaintance with Egyptian hieroglyph and Phoenician and other writings'. From her experience as a teacher, Edna recognized the value of Homer's methods of preserving learning in stories 'put in the form of entertainment, because people learn easily when they enjoy the process'.

When the school at the US base at Burtonwood closed in 1959, Edna was transferred to the London Central High School, for many years located at High Wycombe, where she taught until she retired amid the plaudits of colleagues and students in 1981. After long working days during her years at Central High, Edna would put away her schoolbooks and turn to her research on Homer. She regularly worked until the early hours, and sometimes all through the night; on one occasion she recorded that she had stayed at her desk until dawn broke and had burst into tears when she realized she had made another discovery of great significance.

Edna spent from Monday to Friday each week at her flat in London. On Friday evening she would make the 200-mile train journey home to Bolton, near Manchester, before returning each Sunday evening. Arduous as this regime was, the train journey gave her time to further her thoughts on Homer. In the spring of 1966 Edna's ideas had begun to mature sufficiently for her to prepare notes for a lecture outlining her findings to a class of students. The prompt cards she wrote give one of the few insights into her thoughts on her research:

In the fifth century BC in Greece, during what is called the Golden Age, arts and sciences seemed to spring to life in full maturity, like Athene from the forehead of Zeus. Before this age, however, there was an earlier period of which we know far too little. To understand ourselves, we need to learn about this earlier time. The epic poems of Homer are sources of such information.

We read the epics – the *Iliad* and the *Odyssey* – for narrative, entertainment, philosophy, ethics, history, and so on. Since no man, Homer or any other, lives or has ever lived in a vacuum, untouched by the world about him, we should also expect to find in the epics at least some of the leading ideas about the arts and sciences of his day. Far too little thought has been given to either art or science in Homer, however. Because we live in this highly progressive twentieth century, most of us suppose that we understood more of art and science at the age of ten than anyone living or writing three thousand years ago could possibly have known. We need to re-evaluate such ideas.

I say this because the *Iliad* and the *Odyssey* were the most *thoroughly exasperating* pieces of literature I have ever read or tried to read. I read these two books over and over, again and again. Each time I found the same things: an excellent narrative, superbly told; well-drawn characters; the world's best plots; pathos, horror, excitement, calm; philosophy, history, and so on. Yet, each time I finished reading, my reaction was the same: I felt I had missed the point. I read mythology, I read volumes of ancient history, I went to Greece, I then read criticism and comment by the world's leading scholars in Homeric studies. I reread the *Iliad* and the *Odyssey* several more times, but the same old feeling remained, that I was missing the point. Homer, so it seemed to me, was saying something very clearly, yet something I did not grasp. Between his words and my understanding was a veil.

Eventually, out of all this, emerged a few ideas. Are both epics extended metaphor, from the first words to the last syllable? Figurative language is a poetic device. To sustain a metaphor for the length of two books is long. Nevertheless, I put everything else aside to explore the possibility, for by then I had begun to see what might possibly be the author's purpose. What did this ancient author think sufficiently worthwhile to put into a book-length poem at a time when writing was either unknown or, if known, an expensive process? What was the incentive, good literature apart, for poets and scholars to memorize both books even after writing was known?

Her answer to these questions was: the preservation of astronomical knowledge.

In the 1960s Edna made contact with an academic in an allied field, when she wrote to Dr Giorgio de Santillana at the Massachusetts Institute of Technology after reading his 1961 book *The Origins of Scientific Thought*, which explored how important scientific ideas sprang from the cultural background of their own times. Some letters have survived from this intermittent but lively correspondence with de Santillana and his colleague Dr Hertha von Dechend, of the University of Frankfurt in Germany. In 1969 Dr de Santillana and Dr von Dechend published *Hamlet's Mill*, an influential study on the purpose of myth. In a letter dated 6 May 1966, Edna says:

> I am working with the *Iliad* only at the moment; it appears to be a thoroughgoing study of astronomy, among other things. About fifty-five constellations seem reasonably sure now. Another ten to fifteen seem probable. Two years ago I hoped that five years might be enough time for me to complete what I can of this study . . . Thinking that the originators of what appears to be a fantastic piece of science probably had to do a great deal of their work mentally, I have tried to train myself to project these Homeric skies in like manner. Except for some I have made myself, I have no maps of such early skies. Eventually this study should be finished. When that time comes, if the material seems accurate, I should like to publish.

De Santillana replied within the week: 'We appreciate very much your patient investigation of Homer's astronomy. It is exactly the kind of thing that we are following in various directions . . . you have raised quite a number of hares [and] we are very much hoping for you to help us.' A further, undated, letter that followed this first exchange was signed by both Dr de Santillana and Dr von Dechend:

> . . . von Dechend and I have been obstinately pursuing the same mare's nest that you brought to our attention . . . it gave us immense encouragement to know that you had actually placed a whole lot of these nymphs in the sky where they belong . . . but go and tell it to the philologists. They haven't got either the eyes or the mind to see them. For my part – and hers – we consider your little letters as equivalent to the arrival of an armoured division in the presence of Troy – not to bring any faintheartedness in the souls of those heroes, but just to give them a good spanking and send them to bed! I notice that you quote the works of Carl Blegen, surely a great archaeologist and one who should be on your side if someone told him to stop digging and use his eyes . . . please count on us as a pair of your devoted admirers.

11

The Carl Blegen to whom they were referring was, of course, the famous American archaeologist (1887–1971) who excavated at Troy and discovered what has been called 'Nestor's palace' at Pylos on the Peloponnese, where he found more than a thousand tablets in the ancient Linear B script.

When Edna retired, at the age of sixty-five, her life's work had taken its toll and she was not in the best of health. In her remaining years, she pottered around her home, cultivated her roses, discussed literature and drama with her husband, and spent hours in quiet thought. She also read voraciously from the scores of volumes kept on shelves in her bedroom, but did little if any new work on Homer and was reluctant even to discuss the subject. The enormous task of editing her papers and presenting her work in a typescript would have taken more years, and she could not summon the energy to face that huge endeavour. Until her death, her papers lay almost untouched in a mahogany chest in a corner of the dining room.

Edna left her papers to Florence, who herself retired from teaching in 1992. Then began a two-year process of sorting and cataloguing the 'Leigh Papers', inputting as much text as possible into a computer, and starting to organize it. At this time we had no ambitions beyond the family duty of preserving Edna's lifetime work, and certainly no dreams of ever being able to further it. Our earlier careers had stood at a distance from the world of the classics and astronomy, but, although it at first appeared to pose a number of difficulties, this proved to have significant advantages. Neither of us was burdened with conventional ideas about Homeric epic, and we were able to look at Edna's research with open minds. Florence's experience as a teacher of mathematics was invaluable in understanding the logic of Homeric astronomy. My lifetime career in journalism, as well as jointly writing with Florence two books of regional history which involved dealing with large archives, gave a facility for organizing material in an accessible manner.

Our inquiries into the literature of Homer and the history of astronomy showed that Edna's conclusions, as we had long thought, were without precedent. As the cataloguing came to an end, we had a growing awareness of the significance of the papers, even if the overall astronomical picture could not yet be seen. We made approaches to the academic world with a view to having the work evaluated as a serious study, but several letters to historians of science inviting them to read a summary of Edna's work or see a short presentation of its main points brought polite but disappointing replies.

We then made a number of attempts to finish one or both of Edna's incomplete manuscripts by making use of her scores of pages of handwritten notes. After a break in Connemara in 1994 to gather our thoughts and study this material, we thought we were on the brink of success and returned home in high spirits. Two weeks later we reluctantly had to admit defeat and start all over again. It was becoming apparent that Edna had never committed some of her discoveries to paper. If an unknown amount of material had been lost with her when she died, we feared we would never see her total vision of Homeric astronomy. We abandoned our attempts to create a complete manuscript from her papers and decided to pull together in as logical a fashion as possible all the material we had transcribed. When this was completed we would make another foray into the academic world to seek an evaluation of its worth.

Since beginning the task of cataloguing in 1992 we have made considerable use of modern technology. Lacking Edna's intimate familiarity with the *Iliad*, it would have added years to our work if we had had to search through the epic manually on each of the many occasions when we wanted to compare words and phrases, or to find quotations to cross-check with other translations. We solved this problem by finding a copy of Samuel Butler's 1898 translation on a CD-ROM. So far as we were aware, this was the only translation then available that could be easily loaded into a computer and searched electronically. But, in spite of its usefulness to us, Butler's translation has deficiencies: his style is a prudish reminder that the Victorian Age has long been superseded, and since he wrote in prose he did not feel the need to reproduce all of Homer's poetic epithets such as 'red-haired' Menelaus. Such epithets are important to Homeric astronomy, and had to be checked against a more expansive translation.

It makes little difference which translation is used when reading *Homer's Secret Iliad*, for all translators remain faithful to the sense of the Greek while rendering it into the poetic or prose style of their day. The images the translations convey – whether of battles on Earth or astronomical events in the sky – are constant, and whichever translation suits the reader best is the one to use, whether it has the swashbuckling style of George Chapman (1616), the elegance of Alexander Pope (1715–20), the scholarship of Andrew Lang, Walter Leaf and Ernest Myers (1882), the noble austerity of E. V. Rieu (1950) or the stirring poetry of Robert Fagles (1990).

As well as enabling us swiftly to search the text of the *Iliad*, technology also let us examine on a screen how the skies of Greece would have looked in Homer's era and, indeed, thousands of years earlier. We used a number of astronomical programs, and by inputting a time and the latitude of Athens (38° north), from where the sky was to be viewed, we could see in an instant a chart showing not only the stars and constellations, but also the positions of the Sun, the Moon and the planets for the required year, day, hour and minute. These programs enabled us to check whether the astronomical images we were deriving from the *Iliad* had a basis in astronomical fact.

In the autumn of 1994 I was able to retire early and join Florence in working full-time on the project, and we supplemented our own reading by attending courses on Homeric epic at Manchester University. These proved to be a blessing, for our lecturer, Dr Kathleen Hull, is a scholar of great learning and a teacher of rare talent. Lectures on Homer and Bronze Age Greece by Dr Hull, and later by Ian Howarth of the Open University, led to a deeper understanding of the context of Edna's work, and a study tour of mainland Greece added flesh to the bones of our knowledge. We also began a study course with Linda Simonian at her astronomy centre on the Pennine hills above Todmorden, Yorkshire.

In March 1995 we made another approach to the academic world through Edna Leigh's niece, Dr Peggy Ann Kruger, Director of Public Affairs at the University of Texas. Dr Kruger put us in touch with Dr Betty Sue Flowers, Professor of Western Literature at UT, who agreed to see a presentation of the work during a summer visit to London. Dr Flowers' interest was an important landmark, and in the autumn we spent a month at the University of Texas at Austin, where we gave a presentation to a small group of leading academics. They showed a welcome generosity of spirit, and to our considerable surprise encouraged us to continue our work on Edna's papers. Dr Flowers suggested compiling a page-by-page commentary on the *Iliad*, interpreting as much of the literary narrative as possible from the astronomical angle.

This unexpected encouragement marked the end of a family duty and the beginning of a roller-coaster experience that we could never have anticipated. At a time in our lives when more gentle pursuits beckoned, we began an intensive three years during which we touched unimagined heights and occasionally plummeted to the depths. We soon decided to concentrate our initial efforts on interpretation of the *Iliad*, for within its stirring tales are contained Homer's star and constellation catalogue, without which it is not possible to expound upon other astronomical matters such as the cyclic movement of the heavens and the nature of the universe. The *Odyssey*, the subject of a future book, is more diverse, and without knowledge of the astronomical learning of the *Iliad* it would be difficult to understand.

The creation of an astronomical commentary began slowly, but after a few months we became more attuned to Edna's ideas and interpretation of narrative, and eventually began not only to expand upon these but to make significant advances of our own. At our presentation in Austin, we had said we felt as if we had been looking at a huge box of jigsaw pieces thrown upon the ground. Each separate piece was colourful and interesting, but it was not until the detailed commentary began to take shape that the pieces started to interlock and we at last began to see the glorious expanse of Homeric astronomy. Seven-day weeks and twelve-hour days were usual, and often we worked late into the night, afraid to stop in case the thread of a particular idea was lost.

Our technique in discovering new material was not to identify an astronomical event (such as an eclipse) and then read the *Iliad* to see if Homer might have described it. On the contrary, when we came across a phrase, sentence or paragraph in the *Iliad* that created a mental image of an astronomical event, we turned to the textbooks and computer programs to discover if such an event could have occurred.

In August 1996, and with our work far from complete, there was further academic encouragement during a visit to the University of Frankfurt to meet Dr von Dechend, the colleague of Dr de Santillana. We gave Dr von Dechend and two of her professional colleagues a preview of Homeric astronomy. Our answers to searching questions on various issues must have been satisfactory, for, as we left, she said, 'This is very important. You must publish as soon as possible.' In a later letter Dr von Dechend and Dr Walter Saltzer observed that we had the 'correct amount of obsession' to continue Edna's work.

In 1997 we decided that our work on the *Iliad* should come to an end and began to prepare a manuscript so that others would have the opportunity to enjoy and carry on the remarkable enterprise begun by Edna Leigh. This book is but the beginning of a journey, and as more narrative is interpreted in the years to come others will experience all the excitement and satisfaction of exploring and opening new avenues of Homeric astronomy. Edna believed it would be fifty years before the total substance of Homer's epics would be known. Future researchers may even look beyond the boundaries of astronomy and uncover learning about other matters, such as geometry. But principally our hope is that *Homer's Secret Iliad* will convey extraordinary new insights into the achievements and minds of people who lived three thousand years ago.

1

Astronomy and the Ancients

Astronomy compels the soul to look upwards and leads us from this world to another.

Plato, *The Republic*, 7.528–9

To watch the stars appear in the darkening sky as dusk falls brings a sense of wonder and mystery to even the most casual observer; and so it has been throughout human history. Ancient peoples were as curious about the universe as we are today, and their observations of the skies raised weighty questions to which they sought answers. Scientists now exploring the furthest reaches of space to determine the origins and future of the universe are following directly in the footsteps of those who created myths and religions to comprehend the heavens and to explain the creation of the world.

The night skies would have been much more familiar to the ancient Greeks than they are in our times, when the spectacle of the heavens is much reduced by the pollution of artificial lighting. Many fainter stars that could be seen with the naked eye in Homer's times are no longer visible to city dwellers, their pinpoints of light swamped by the glare from street lamps and floodlit buildings. Not only has the spiritual impact of the skies largely been lost, but such practical matters as the calculation of the hours, days, months, seasons and years are now more easily handled by electronic clocks and printed calendars. Nor do we need to understand Homer's great astronomical theories by sitting outside in the dark as a poet-singer recites from the epic and points towards the heavens to show how the stories of the Siege of Troy become allegories for identifying stars and describing the movement of the heavens. Such ideas today are conveyed in books and on computer screens.

To understand the movement of the heavens, it is easiest to imagine the Earth as being at the centre of a huge celestial sphere that rotates around us once each day. This is how the universe appears to an observer on Earth, and it was a model generally accepted until Copernicus showed in the sixteenth century AD that in fact the Earth and other planets orbit the Sun. In the ancient concept, the stars maintain their positions relative to each other and so appear to be fixed on to the sphere, whereas the Sun, the Moon and the planets have individual paths through the sky. At any one time, an observer can see only half of the celestial sphere, the other half being below the horizon.

Every twenty-four hours the stars visible from a particular latitude pass across the sky, but the light from those in the sky during the day is overwhelmed by the glare of the Sun. As the seasons of the year progress, so does the panorama of the skies, and stars visible at one time of year at night are at other times in the daytime sky. Only at the Equator can the entire celestial sphere be seen (see fig. 4).

Newcomers to stargazing today face many of the same problems as did their ancient ancestors. There can be no understanding of the skies until the multitude of stars has been divided into groups and so organized into a comprehensible form. Observers of thousands of years ago solved this problem with great imagination and created the patterns known as constellations by joining stars together by means of imaginary

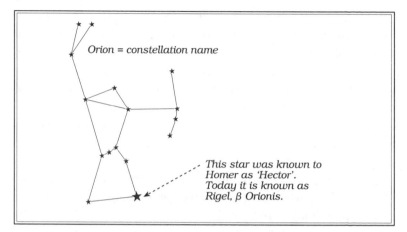

Fig. 1 Identifying stars

lines. Many of the constellations we recognize today had their origins in those times. Constellations are often associated with mythological characters, animals and fish, as well as inanimate objects and abstract shapes such as Lyra (the Lyre) and Triangulum (the Triangle). The reasoning behind constellations such as Orion (the Hunter – fig. 1), Gemini (the Twins) and Leo (the Lion) seems quite obvious. Others, like Auriga (the Charioteer) and Aquarius (the Water Carrier), severely test the observer's imagination, but it will be shown in Chapter 9 how Homeric geography can account for those patterns that seem to have little in common with their names.

Individual stars in constellations also have to be identified, and this has been done in a variety of ways in differing cultures. Homer, for instance, identifies most if not all of the few hundred brighter stars as warriors, and the more powerful the warrior, the brighter his 'personal' star. Ancient names for stars in other cultures have survived to this day, and include Sirius, Antares, Spica, Aldebaran and Regulus – respectively, the brightest stars in the constellations of Canis Major, Scorpius, Virgo, Taurus and Leo.

In modern times the brighter stars in constellations are generally identified using the system laid down in the seventeenth-century catalogue of Johann Bayer, who labelled them in order of brightness with the letters of the Greek alphabet (see Appendix 1). The brightest star is usually described as α (alpha), the second brightest as β (beta), and so on.

The number of stars in a large constellation can soon surpass the number of letters in the Greek alphabet, and a certain amount of ingenious duplication is used. In Orion's bow, for instance, a line of stars is labelled π^1, π^2, π^3, π^4, π^5, π^6 Orionis (where 'Orionis' means 'of Orion'). There are also some discrepancies – for example, in Orion the brightest star is denoted as β, and the second brightest as α. To overcome these limitations, John Flamsteed, the first Astronomer Royal, listed more than 3,000 stars by number in his catalogue *Historia Coelestis Britannica*, published posthumously in 1725. The stars π^1, π^2, π^3, π^4, π^5 and π^6 in Orion's bow are 7 Orionis, 2 Orionis, 1 Orionis, 3 Orionis, 8 Orionis and 10 Orionis in Flamsteed's catalogue. In this book, stars may be identified by both modern and Homeric labels. 'Hector, Rigel, β Orionis' gives the star's Homeric name (Hector) followed by its common name and then by its Bayer catalogue definition.

The brightness of stars can also be quantified, and the modern method uses a logarithmic scale of magnitude in which the higher the number, the fainter the star. Most of the stars visible to the naked eye have a magnitude between 0 and 6. A star of magnitude 1 is 100 times brighter than a star of magnitude 6. The very brightest stars may even have negative magnitudes – the brightest star of all, Sirius, α Canis Majoris, which far outshines all others, has a magnitude of −1.4. A 'v' after a magnitude indicates a star that varies in brightness. How Homer compared the brightness of stars is shown in Chapter 4.

An impressive feature of the heavens is the Milky Way, the light from the uncountable stars of the galaxy of which Earth is a part. The brilliant streaks of meteor showers flashing across the sky, the rare burst of light from a supernova or exploding star, and the strange wanderings of planets are other events seen by ancient peoples and recorded by Homer.

The zodiac is a band of sky some sixteen degrees wide that appears to encircle the Earth and to contain the paths of the Sun, the Moon and the planets. It is generally thought that the concept of the zodiac originated in Mesopotamia, but it is not clear when, nor who formulated it. The Egyptians were aware, too, that within the zodiac were the paths of the Sun, Moon and the planets, and their learning may also have originated in Mesopotamia.

The attention of ancient peoples would have been drawn to the zodiac for a number of reasons. They would have noticed that, although most of the stars seemed to be fixed to the celestial sphere, their positions remaining fixed relative to each other, five 'stars' moved independently against this background of fixed stars. The five wanderers were in fact the planets visible to the naked eye that we now know as Mercury, Venus, Mars, Jupiter and Saturn. The Sun appears to travel along the centre of the zodiac, and its apparent path through the sky is known as the ecliptic (fig. 2). The path of the Moon is also confined to the zodiac, and observations of the Sun and Moon became important in calculating time as astronomical knowledge grew.

The zodiac is divided into twelve divisions, each of thirty degrees. Each division is associated with a constellation, and, although the zodiacal constellations of Aries, Pisces, Aquarius, Capricornus, Sagittarius, Scorpius, Libra, Virgo, Leo, Cancer, Gemini and Taurus may be most commonly known for their connections with astrology, their primary purpose is astronomical. As the Earth rotates on its axis, the Sun is said to rise in a specific constellation as it comes into view – i.e. just before dawn a particular constellation is visible on the horizon at the point where the Sun is about to rise (this constellation is Taurus in fig. 3). As the Earth is also orbiting around the Sun, the

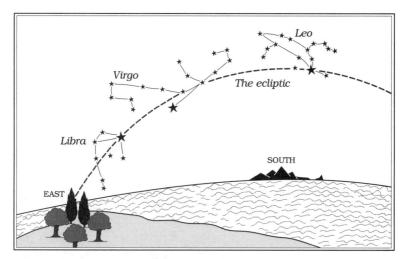

Fig. 2 This illustration shows three of the twelve constellations of the Zodiac. As they move in an arc from the east and set in the west, they, and the constellations that precede and follow them, give the impression of rotating around the Earth. The zodiacal band extends about eight degrees on either side of the ecliptic, the line (dotted on the diagram) that marks the apparent path of the Sun during the day.

constellation in which the Sun rises changes during the year from Taurus to Gemini etc. The Sun spends about one month each year in each constellation, and when a cycle of it rising in all twelve zodiacal constellations is complete, a solar or tropical year has passed. As we shall see, Homer used zodiacal constellations in allegories to explain the movement of the heavens over thousands of years.

The Moon has a very different cycle, there being an interval of 29½ days between one new moon and the next, and twelve of these lunar months make up a lunar year of 354 days. The lunar year and the solar year are quite independent, and when the first calendars were created there were problems in reconciling the lunar calendar with the solar calendar.

Of great importance to travellers and navigators since ancient times has been the north celestial pole, the imaginary

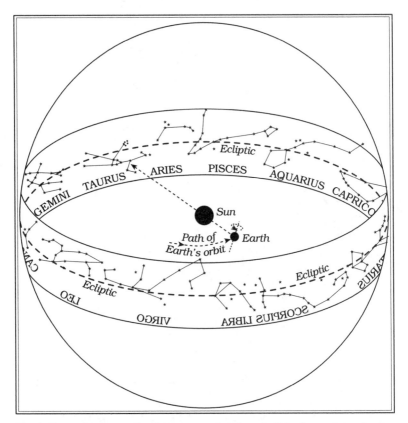

Fig. 3 *To an observer on Earth it appears that the celestial sphere rotates about us once each day, owing to the spin of the Earth. The Earth also orbits the Sun each year, and the Sun appears to move around the zodiac. At the season of the year shown in this diagram the Sun would rise against the background of the stars of Taurus, but as the Earth moved on it would next rise against the background of the stars of Gemini. After the Sun had passed through each of the twelve zodiacal constellations a year would have passed.*

point at which the Earth's axis of rotation, if extended, would meet the celestial sphere. As fig. 4 shows, the entire heavens appear to rotate around this point, and the star closest to the celestial north pole at any one time is known as the pole star and can be used as a fixed reference point for travellers.

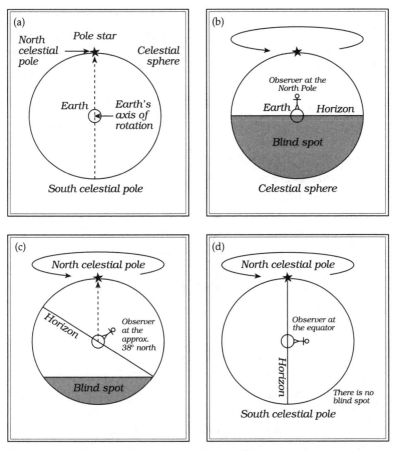

Fig. 4 *The heavens as seen from various places on Earth.*
(a) The celestial poles are the points where the Earth's axis of rotation, if extended, would meet the celestial sphere. The star nearest the north celestial pole is known as the pole star.
(b) At the North Pole, the same area of the celestial sphere is above the horizon at all times and the visible stars never rise or set – they are all circumpolar, appearing to rotate about the north celestial pole. Half of the sky is never visible, in the 'blind spot'.
(c) At latitude 38° north the area visible above the horizon changes as the Earth rotates, though some stars always remain in the blind spot and circumpolar stars always rotate around the north celestial pole.
(d) At the Equator there is no blind spot: all the visible stars come into view as the Earth rotates.

Because the Earth slowly wobbles like a spinning top as it orbits the Sun, the point at which its axis would meet the celestial sphere varies, so the same star does not always indicate celestial north. Thuban, a star in the constellation of Draco, which was the pole star five thousand years ago, is now a great distance across the sky from the north celestial pole. The changes of pole star are slow, but eventually the pole star of our own times, Polaris, in the constellation of Ursa Minor (the Little Bear), will also be replaced by another star on what is known as the precessional circle. Although he almost certainly did not know why they happened, Homer was aware not only of these changes to the pole star, but also of others caused by the wobbling of the Earth's axis.

In the course of a night, many constellations can be seen rising in the east and setting in the west, but a number rotate around the pole star and never set. These are known as circumpolar constellations.

Astronomy is the oldest of the sciences, and has been associated with myth and religion throughout history. The megalithic monuments that include Stonehenge and scores of other groups of standing stones in Britain and Europe, inscriptions on clay tablets from Mesopotamia, and the temples and pyramids of Egypt are evidence of astronomical cultures that had their origins before 3000 BC. Whatever the religious demands, there were also sound practical reasons for making and preserving astronomical observations. Perhaps the most important were marking the passage of time, the making of calendars and the use of the stars for navigation.

Recording the passage of time goes back far into human history – long even before the invention of the wheel. An engraved bone found in South Africa, dated to around 37,000 BC, has been said to resemble the wooden calendar sticks still used by bushmen in modern times. Another engraved bone

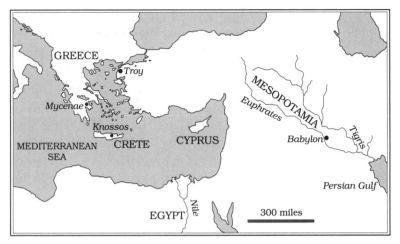

Fig. 5 *Three of the world's earliest astronomical cultures arose in Mesopotamia, Egypt and Greece.*

about 12,000 years old was found in a cave in France, and it has been suggested it could have been a solar calendar.

Knowing the time of day and year was as important in ancient times as it is today, but, while we mark the passing of our lives with watches on our wrists and calendars pinned to a wall, the ancients studied the heavens. The hours of day were determined by the position of the Sun, and the hours of night by the passage of constellations across the heavens. In the *Iliad*, Homer marks the passage of time by the Sun's position in the sky: 'Thus all day long the young men worshipped the god . . . but when the sun went down, and it came on dark, they laid themselves down to sleep by the stern cables of the ship, and when the child of morning, rosy-fingered Dawn, appeared they again set sail' (1.472); 'Now so long as the day waxed and it was still morning, their darts rained thick on one another and the people perished, but as the hour drew nigh when a woodman working in some mountain forest will get his midday meal . . .' (11.84); 'So long as the sun was still high in mid-heaven the weapons of either side were alike deadly, and

27

the people fell' (16.777). Similarly, when the warriors Odysseus and Diomedes set off at night to raid the Trojan lines, Homer says, 'The stars have gone forward, two-thirds of the night are already spent, and the third is alone left us' (10.251).

The beginning of a year could be marked by the first annual sighting just before sunrise of a bright star that had previously been in the daytime sky and so obscured by the light of the Sun. This sighting is called a heliacal rising, and when that star was next first seen just before dawn a year had passed. From before 3000 BC, the heliacal rising of Sirius, the brightest star, was the basis of an Egyptian calendar.

The ancients observed that twice each year, when the Sun rose at due east and set at due west, the hours of day and night were equal. These times are known as the vernal (spring) and autumnal equinoxes. Two other important annual solar events were also noted – the longest day (the summer solstice) and the shortest day (the winter solstice). Evidence of prehistoric learning about the equinoxes and solstices is sparse, but in 1998 it was announced that the oldest known stone monument believed to have an astronomical purpose, dating from 4500–4000 BC, had been found at Nabta, a site on the edge of a dried-out lake in the western desert of southern Egypt.[1] A stone circle with four sets of vertical sighting stones indicated the summer solstice when the Sun was directly overhead. Two sets of these stones are aligned in a north–south direction, and a second pair gives a line of sight towards the point on the horizon where the Sun rises at the summer solstice. About a mile away from the circle there is an east–west alignment between one megalithic structure and two others. Two more lines of twelve stones lead north-east and south-east, but their significance is not yet known. The site was abandoned after the lake dried up c. 2800 BC, but Professor J. McKim Malville of the University of Colorado believes the people who built Nabta may have helped dynastic culture to evolve and hastened development of the pyramids.[2]

By using the skies to determine the time of year, farmers could plan their sowing and harvesting. Without a reasonably accurate indication of the season, it would have been possible in northern climes to mistake a warm spell in winter for the advent of spring and plant seed only to have it rot in the ground. Conversely, planting delayed too long would leave crops lying unripened in the fields when autumn came, and bring the spectre of hunger in the following winter. In warmer regions, the changing skies would have heralded the onset of wet and dry seasons – flood and drought. Hunters and fishermen could relate the passing of the months in the heavens to the annual migration patterns of their quarry.

More precise observations would have been needed to celebrate religious festivals associated with the equinoxes and solstices, and no doubt the calendar of the heavens would have marked the times to pay taxes. Mothers would have known in which of the twelve constellation of the zodiac the Sun was rising when their children were born, and would have recorded the passing of a birthday when the Sun rose the following year in the same constellation.

Not only do simple observations of where the Sun rises and sets on the horizon at the equinoxes mark east and west, but the Sun at its highest daily point in the sky denotes south, while north is easily found at night by the pole star. Such knowledge of the points of the compass would have been vital for trade across land and seas between widely separated countries. The world's oldest shipwreck,[3] a sixty-foot sailing ship dated to 1316 BC, found off the south-west coast of Turkey, confirms that pre-Homeric peoples of the Aegean had the vessels and knowledge of navigation and geography to trade over wide areas. Among the goods found in the wreck were ebony, ostrich eggs and hippopotamus teeth from Africa, swords from Italy, tin possibly from Afghanistan, and items from Egypt. Other Bronze Age wrecks found off Turkey confirm a thriving foreign trade, and their crews, like Homer, would have looked to the stars for guidance.

Though the hunter-gatherers of the Stone Age were early observers of the heavens, the science of astronomy would have gathered pace from necessity when men first came together to farm crops and domesticate and breed animals in small communities in the fertile river valleys of Mesopotamia, Egypt, China and India. At about 8000 BC, the settlement of Jericho in the Jordan valley was well established and surrounded by a massive stone wall, and cereals were being grown on the banks of the river Nile in Egypt by at least 5000 BC.

In fertile land by the Euphrates and Tigris rivers in Mesopotamia – modern Iraq – farming was so advanced at the end of the seventh millennium BC that irrigation channels had been built to water fields. Small communities there grew into the city states which by *c.* 2370 BC had been consolidated into an empire ruled by King Sargon of Sumer and Akkad. The oldest account of the nature of the universe yet discovered in Mesopotamia was inscribed on clay tablets during the Third Dynasty of Ur, *c.* 2113 to 1991 BC, but is thought to have been the record of much older learning.

Astronomical evidence of that region is scant before the first Babylonian empire succeeded the Sumerians at the beginning of the second millennium, but such material and mythological stories as have survived indicate an awareness of the heavens. The Sumerians invented a system of writing *c.* 3100 BC, and incorporated astronomical concepts in two of their characters. One of these represents a star, and the other shows the 'sun over the horizon'. They are also said to have arranged groups of stars into constellations and to have named at least some of them after the same things by which they are identified even today. These include the Bull (Taurus), the Lion (Leo) and the Scorpion (Scorpius). This achievement alone would indicate that their observations were beyond the trivial or superstitious. The Sumerian *Epic of Gilgamesh* was found on cuneiform tablets believed to have been written at about 1800 BC but describes a legendary hero of at least a thousand years earlier and contains images which have been associated with the heavens.

One of the best-known studies of how stars and constellations came to be recognized is Richard Hinckley Allen's *Star Names, Their Lore and Meanings*, first published in 1899 and still in print today. Allen records that Scorpius was known in Mesopotamia at *c.* 5000 BC, Cancer at *c.* 4000 BC, and Taurus at *c.* 5000 BC.[4] Even so, these dates are unlikely to be those when constellations were first devised.

From Mesopotamia in the first half of the third millennium BC, a lunisolar calendar, in which months are calculated according to the cycle of the Moon but years are solar, is an indicator of astronomical observations being used for practical purposes. The solar year was first divided into two seasons, but by 2400 BC a calendar of twelve months, each of thirty days, was in use. Three centuries later a lunar calendar became the preferred method of marking time. The lunar year of 354 days and the solar year of 365¼ days are independent of each other, and to periodically bring them back into line an extra month was inserted from time to time into the lunar calendar. With the rise of Babylon in the early second millennium, the lunar calendar remained in use until the decline of that empire long after the time of Homer.

For the Babylonians, there was no division between observing, recording and predicting natural celestial events (the science of astronomy) and the use of observations as omens (the pseudo-science of astrology). Divination was a considerable influence, and involved more than just the study of the skies: even the examination of animal entrails and the flights of birds could form the basis of predictions. The astrologers or priests who scoured the heavens for signs which they believed might affect the well-being of their kings and their nations made their forecasts from astronomical observations over many centuries. It was the greatest good fortune when archaeologists found seventy inscribed clay tablets on which were written 7,000 omens that were said to have been sent by the gods as messages of warning to a king or to tell him of their approval. These tablets date from about the beginning of the first millennium, but some of the omens preserved on them might have been made as early as the third millennium. For astronomers, perhaps the best-known tablet is a copy of one first written in the reign (1646–1626 BC) of Ammaziduga, giving the heliacal risings and settings of Venus over a period of twenty-one years. Such records indicate a highly developed system of observations, and are almost certainly part of a much wider body of learning that has been lost.

The association of mathematics with astronomy is very ancient, and yet another great achievement of the peoples of Mesopotamia was the devising of the sexagesimal numerical system, based on a scale of 60, invaluable when calculating the 360 degrees of the circle, angles, degrees of arc and the passage of time in seconds, minutes and hours. The circle was so conveniently divided into 360 degrees because of the number of days used to mark the apparent journey of the Sun around the Earth. An extra five days were added to make up the year to 365 days.

In the centuries and millennia before Homer, there was also the great civilization of Egypt, where events observed in the skies were important for both practical and religious purposes. In ancient Egypt, some groups of stars were configured in a similar way to those of Mesopotamia but were seen as different characters. Draco, the Dragon, became the Crocodile, while Sagittarius is recorded at the fourth-century-BC temple of Dendera on the river Nile as a lion-faced archer. The stars of Orion are represented at Dendera, but Allen said they had also been carved on to the walls of a temple at Thebes as early as 3285 BC. The constellation Aries as a ram was identified with the god Amon, and tombs of Egyptian royalty from the second millennium include paintings of constellations.[5]

The creation myths and religion of Egypt have powerful associations with the skies, and worship of the Sun-god, Amon-Ra, was a central theme. The alignments of temples and shafts and tunnels in pyramids with significant stars and sunrise at the solstices has long been discussed, and particular attention has been given to Thuban, the star closest to the north celestial pole from about 4000 to 1800 BC. Thuban is also important in Homeric astronomy, and lies at the heart of the *Iliad* and early Greek ideas on the slow but constant apparent movement of the heavens (see Chapter 7).

The calculation of time was a skill acquired early in Egypt, and a lunar calendar was in use even before the pyramids of Giza were built in the third millennium. The heliacal rising of Sirius marked the beginning of a new year, and coincidentally this event occurred shortly before the annual flooding of the Nile that spread life-giving mud on the fields. The year was divided into twelve lunar months (354 days), and, as in Mesopotamia, to keep the calendar in step with the solar year of 365¼ days, an extra lunar month was periodically added to the lunar year.

Although adequate for marking religious purposes, this calendar did not meet the requirements of an increasingly sophisticated society, and in the third millennium a calendar of 365 days – a quarter of a day short of a solar year – was created for bureaucratic purposes. The new calendar ran alongside the old, and was based on a solar year divided into twelve months each of thirty days, with five days added at the end of the year. There was no accommodation of leap years in this new civil calendar, which became increasingly out of alignment with the seasons until, after some 1,460 years, the heliacal rising of Sirius again occurred on the first day of the calendar year. This period was known as the Sothic cycle, after the god Sothis, who represented Sirius. To overcome the discrepancies between these calendars, a second lunar calendar was introduced for religious use, based on the civil calendar rather than the rising of Sirius. Nevertheless, the first lunar calendar was not forgotten and continued in use for agricultural purposes. Such were the complexities of calculating time in ancient Egypt.

Our modern calendar owes its origins to the Egyptians, as does the division of the day into two periods of twelve hours each, whose passage was recorded by shadow clocks by day and water clocks at night. As well as the heavens being crucial to their religious beliefs and the pursuit of agriculture, a knowledge of navigation, as likely as not based on the stars, was required for the Egyptians to trade with the wider world and make war in foreign lands. On the basis of what has so far been discovered, the astronomical achievements of the Egyptians have not been as highly regarded as those of their contemporaries in Mesopotamia, but the whole story may not yet have been told. E. C. Krupp, an American historian of astronomy, has said we might yet be surprised at what will be found among the tombs and temples of the Nile.[6] The Egyptian influence on the calendar and the twenty-four-hour day, together with the Babylonian sexagesimal system, illustrate how ideas from those ancient cultures still affect modern time-keeping.

Homeric astronomy reflects influences from the Middle East and Egypt, but elsewhere in the world, from the pre-Columbian Americas to ancient China, other civilizations were making observations and developing their own astronomical cultures. The creation of calendars and the careful, if mis-guided, examination of the heavens for omens believed to affect life on Earth have been common preoccupations.

Chinese astronomy was in existence long before 2000 BC, and by the middle of the second millennium the Chinese had a cal-endar that defined the solar year at 365¼ days and the lunar month at 29½ days. Written records from later centuries note such events as meteor showers, and Chapter 6 will indicate that one rare event seen in the Far East may also have been observed in Greece and recorded in the *Iliad*. The skies were charted from early times, and astrological significance was read into observations of unusual events such as solar and lunar eclipses.

'Obsession' is a word that has been used by several researchers investigating the creation of a calendar of great accuracy by the Mayans of Meso-America. Their civilization was at its height in *c*. AD 250–900 and, although little remains of their hieroglyphic writing, a remarkable document known as the Dresden Codex contains astronomical calculations, including a calendar based on Venus that was accurate to one day in 500 years.[7] Other documents reveal mathematical skills that enabled the Mayans to record astronomical cycles over long periods and to forecast solar eclipses. The Mayan calendar was used both for civil purposes and for the timing of rituals in which it is believed the Sun and Moon were exalted and prisoners of war were cruelly tortured and killed to placate the gods.

The Aztec empire of Mexico, crushed by the Spanish in the sixteenth century, was also pervaded by harrowing religious ritual and sacrifice in ceremonies honouring the Sun-god. The Aztec calendar was derived from that of the Mayans, and had ritual and civil cycles that came together every fifty-two years – an occasion for much bloodletting.

At Nazca in Peru, alignments of white stones on the ground trace huge figures such as a bird, a monkey, flowers and geometric designs. These can be seen when viewed from the air. The origin of the figures is a mystery, but they were created before the rise of the Incas in the twelfth century,[8] and some investigators have given them astronomical connotations.

Astronomical learning was put to practical use by the Polynesians, who used the stars to voyage with great accuracy over vast distances out of sight of land. Their ancestors had moved into the great southern ocean before the Christian era, and it is not surprising that the creation myths of these nautical peoples encompassed the sea, the sky and the Earth.

Northern Europe too has evidence of astronomical learning from the period before the third millennium, as many groups of standing stones in Britain and Europe testify. The purpose of the megalithic monument at Stonehenge arouses great passion, with some believing the monument had a considerable astronomical purpose and others being dismissive of such suggestions. In his controversial work *Stonehenge Decoded*,[9] Gerald Hawkins claimed that the stones were a sophisticated prehistoric observatory, but others have been more cautious in their views. Nevertheless, a number of alignments of the stones indicate astronomical reference points, the best known being that of the Heel Stone with sunrise at midsummer. Also, the open end of a horseshoe of standing stones and the impressive processional avenue are aligned to the north-east, the point on the horizon where the sun rose at the summer solstices thousands of years ago.

The first of three stages of development at Stonehenge began before 3000 BC and predates the building of the earliest pyramids. The site then continued to be developed and remodelled until the second millennium. There is no reason to believe that the peoples who built and extended Stonehenge for a thousand years and more were any less dependent on the stars for marking time and the agricultural seasons, and for travel and religion, than their contemporaries elsewhere. Nor is there any obvious reason why a human concern identified in early Mesopotamia, Egypt, China and South America should not also have been present in Europe. There is also no apparent reason to discount basic astronomical learning being inextricably associated with the elaborate religious beliefs and rituals sometimes postulated for Stonehenge. In his exhaustive investigation into Stonehenge, John North is persuasive when he says, 'The stones were not erected as a means to *investigating* the heavens in a detached and abstract way. The aim was not to discover the patterns of behaviour of the Sun, Moon or stars but to *embody* those patterns, already known in broad outline, in a religious architecture.'[10]

Until the rediscovery of Homeric astronomy, the history of astronomy in Greece has always begun with Thales of Miletus (625–547 BC), often known as the 'Father of Greek Science'. Little is known of Thales' work, and none of his writings has survived, but he is said to have created a cosmology in which water was the essence of all matter. A traditional belief that he forecast an eclipse of the Sun in May 585 BC is disputed by modern scholars.

Before Thales, there are references to the stars and constellations in the works of both Homer and his contemporary Hesiod. Homer briefly declares a familiarity with a number of constellations and groups of stars, and names Orion, the Great Bear (Ursa Major), the Wain (the Plough or Big Dipper in Ursa Major), Boötes, the Pleiades and the Hyades in Taurus (18.486), as well as Sirius, the brightest star (22.29). Sailing instructions given to Odysseus (*Odyssey* 5.271) are stated with such confidence that they must reflect the learning of a man who could use the skies for practical purposes and who had more than a passing acquaintance with the heavens. It is not realistic to take the view that the naming of a handful of stars and constellations could have been the total astronomical learning of Homer's time and that all of the other stars in the wide expanses of the skies would have been ignored. The Homeric astronomy explored in later chapters indicates that his knowledge of the heavens was much greater than this.

In *Theogony*, Hesiod wrote of the creation of the world in a mythological manner, but he adopted a more practical approach in *Works and Days*, an almanac for the farming year. In this he gave advice based on observations of the heavens, and used the heliacal risings and settings of constellations to instruct farmers when to plough and harvest their crops:

> When the Pleiads, Atlas' daughters, start to rise
> Begin your harvest; plough when they go down.[11]

This familiarity with the skies raises the questions of from where Homer and Hesiod obtained their knowledge, how it had been preserved, and for how long it had been known. That the information they recorded in their poems was in common circulation is evident from the fact that Homer directed it towards travellers and Hesiod towards farmers and the agricultural tasks that had to be performed at the various seasons of the year. Additionally, it must be asked why, when we know of astronomical learning in contemporary Babylon and Egypt, almost nothing has been known of astronomy in Greece, Asia Minor and Crete before Homer's era. The abundance of astronomical learning we have rediscovered in the *Iliad* shows that earlier Greek peoples did have astronomical knowledge, but it was preserved not in writing but through the medium of allegory, using myth and legend.

In recent years there have been increased indications of a knowledge of astronomy by the seafaring and cultured Minoans of Crete, who from sometime before 2200 BC created the first civilization of the Aegean region. By *c.* 1600 BC the Minoans began to be succeeded in power and influence by their neighbours, the Mycenaeans of the Greek mainland. The Mycenaean civilization provides the historic background to Homer's epics.

Until the advent of Homeric astronomy, little was known of the accumulated learning of the Minoans and Mycenaeans before the time of Homer. Investigations in the nineteenth century into the alignments of Cretan temples with the rising sun were not conclusive, but recently the picture has begun to change. Research by Göran Henriksson and Mary Blomberg of the University of Uppsala in Sweden[12] has indicated that the Minoans of Crete conducted systematic observations of the Sun, Moon and stars, and used them to navigate and to regulate their calendar as well as for ritual purposes. Henriksson and Blomberg say their findings make a case for the discovery by the Minoans of the Middle Bronze Age of the *octaëteris*, an eight-year cycle of the Sun and Moon that brought the solar and lunar calendars back into relative harmony. Ancient observers had noted that in eight solar years, each of 365 days, there are 2,920 days. During the same period there are 99 lunations, each of 29½ days, giving a total of 2,920½ days – an apparent difference of only a half-day between the solar and lunar calendars. By the fourth century BC, the solar year was accepted as 365¼ days, which over eight years gave a total of 2,922 days, making the *octaëteris* less accurate than had been thought. Indications that the *octaëteris* was known in Greece over such a long period imply a continuity of learning from the Minoan era to post-Homeric Greece.

A second thread of astronomical learning, reaching from the third millennium to long after Homer, is revealed in a poem written by Aratus in 270 BC in honour of the mathematician and astronomer Eudoxus of Cnidus, who lived in the fourth century BC. In the *Phaenomena*, Aratus put into verse an astronomical treatise by Eudoxus and named forty-five constellations and a number of stars. Curiously, the position in the skies of constellations described in the *Phaenomena* are those not of the times of Aratus and Eudoxus, but of a much earlier era. In a paper on the origins of the constellations, Michael Ovenden of Glasgow University postulated that the skies described by Aratus could have been seen between 3000 BC and 1800 BC at a latitude between 34½° and 37½° north – a band that includes Babylon and Crete. The constellations, he said, were not the work of primitive people but an ordered mapping of the skies, and could be used as an agricultural calendar as well as for navigation at sea. The *Phaenomena* of Aratus was, he declared, a manual in poetic form that enabled seamen to use the stars for navigation. Of the contenders for the title of constellation-makers, he put forward the claim of the seafaring Minoans of Crete – a people who lived not only at the right time but also at the right latitude. Addressing the question of how learning could have survived from Minoan times to the era of Eudoxus and Aratus, Ovenden said, 'the tradition of the constellations must have been handed down by word of mouth for two millennia'.[13] As he wrote those words in 1965, Edna Leigh was independently determining exactly how the oral tradition had been used to preserve astronomical learning in the *Iliad*.

A further paper on the origins of the constellations, by Archie E. Roy, a colleague of Ovenden, states that detail from the *Phaenomena* leaves little doubt that Aratus' poem refers to a 'celestial sphere about 2000 BC with an uncertainty of 200 years each way'.[14] He agrees with Ovenden on the astronomical significance of the Minoans, but points out that the Sumero-Akkadians from Mesopotamia were using a system of constellations 'similar to that given in Aratus as far back as 2100 BC and probably for many centuries before that date . . . perhaps we should assess their [the Minoans'] fitness not to be the constellation-makers in the sense that the Sumero-Akkadians were, but to be the people who took over and modified the Euphratean system for their own navigational use'.

Other obvious questions arising from *Phaenomena* are, From where did Eudoxus obtain his out-of-date information about the skies, and why did he record stars that *could not* be seen from where he lived in his times, and yet not record stars that *could* be seen? The source of his ancient information has traditionally been said to have been a celestial globe brought back 'from Egypt' and on which were carved or inscribed the stars and constellations later described in such detail by Aratus. Eudoxus is an astronomer of repute who not only introduced geometry to the study of the skies but gave the first explanations of the motions of the Sun, Moon and planets. It would seem strange if a man of such achievement was unaware that the information on the globe did not match the skies he knew from his own observations. If, however, he knew that the globe was the repository of ancient Greek observations that were still revered as folklore, it would support the proposition that Homer, too, preserved very ancient learning in the *Iliad*. When it is realized to what extent astronomical matters pervaded the earliest civilizations in every corner of the earth, long before his time, and the level of sophistication brought to the study of the stars for whatever reasons, it immediately becomes more reasonable that Homer too should have concerned himself deeply in them.

The intention in the latter part of this chapter has been to provide such evidence as there is for the case that the astronomical knowledge contained in the *Iliad* should be seen as part of a continuous thread of astronomical learning passed down from Minoan and Mycenaean times to long after Homer. It would be rare chance indeed if Eudoxus and Homer each independently preserved knowledge of the skies from about the same period of the third millennium BC. The possibility of a calendar system known in Minoan times remaining in use in Greece until after Homer adds another strand to the idea. Homer may have been the master poet-scientist who wove a remarkable amount of ancient learning into an epic about the Siege of Troy, but he himself was only a staging post in a long tradition of preserving learning in the oral tradition. Unsung are the many bards who came before him, and their achievements will never be known except through the epics of Homer. In the following chapters, as the story of how Homer uses narrative for this purpose unfolds, it will be seen that the learning of Homer and his predecessors was not a haphazard collection of scraps of knowledge, but a breathtaking construction of immense skill and content.

2

Preserving Learning in Epic

I believe Homer's function was to preserve
knowledge first and foremost, as had a long
line of bards before him.

Edna Leigh

The abundance of astronomical learning in the *Iliad*, which
this book seeks to illustrate, raises the question of why
epic was the medium in which it was preserved. The answer
is to be found in the cultural development of the two great
civilizations of the Minoans of Crete and the Mycenaeans of
the Greek mainland. Like contemporary Babylonians and
Egyptians, the Minoans lived in a rich, powerful and cultured society, whose trading ships plied the Mediterranean.
Ample evidence of their artistic and creative talents in the
forms of elegant pottery, jewellery and beautiful frescos has
been uncovered by archaeologists in the ruins of a monu-
mental palace at Knossos, at Phaistos and at other sites in
Crete.

In 1899 the British archaeologist Sir Arthur Evans began
excavations at Knossos and uncovered a hitherto unknown
civilization. The palace at Knossos is more than five acres in
extent, and its maze of passages brings to mind the myth of the
Minotaur and the labyrinth of King Minos. And, indeed, it was

after that royal figure that Evans gave the name 'Minoan' to the civilization he had discovered.

By 1400 BC the Minoan civilization had declined and the centre of power in the Aegean moved to mainland Greece, where imposing monuments of the succeeding Mycenaean empire can be seen in the fortresses of Mycenae and Tiryns, in the magnificent palace at Pylos and at other locations. So huge were the stones used to build Mycenae and Tiryns that later generations believed it was beyond the abilities of mortal men to move them, and attributed the construction of the strongholds to the Cyclopes, a mythical race of giants. The Mycenaeans had begun their ascendancy at the beginning of the Late Bronze Age, c. 1600 BC, and their influence extended to Crete and Cyprus, but by c. 1100 BC their time too had passed and they faded from the scene. The Greek mainland and islands then passed into a dark age in the sense that almost nothing is known about it. As a historian of Greece, Peter Green, says:

> Before Homer and Hesiod, to put it bluntly, we lack the entire intellectual framework of Greek society, and the far-ranging archeological discoveries made during the past 100 years should never obscure this central truth . . . Unless we count the oldest layer . . . in Homer and other accounts of the traditional myths, we have no literature whatsoever to illuminate either Crete or Mycenae. Oddly, this is not true of other Near East civilizations.[1]

While the Minoans and Mycenaeans were capable of triumphs in the plastic arts, little if anything is known of their myths. Among the important discoveries made at Knossos were numerous clay tablets inscribed with two unknown scripts. Sir Arthur Evans labelled these scripts Linear A and Linear B, but they defied all his attempts to decipher them. Later, tablets inscribed with Linear B were found in greater quantity on the Greek mainland at Mycenae, Pylos and Thebes. Linear A was in use in Crete from about the eighteenth century BC to the fifteenth century BC and is still undeciphered. Linear B is said to be a form of that script modified by the Mycenaeans, and examples have been found on vases and clay tablets dating from about 1400 BC to approximately 1200 BC.

In 1936 Evans gave a lecture on the tablets to an audience that included Michael Ventris, a schoolboy who later became an architect. Inspired by what he heard, Ventris spent years working on the strange characters inscribed on Linear B tablets, and in 1952 he announced that they had been deciphered. Great indeed was the surprise of classicists and archaeologists when Ventris revealed that Linear B, the language of the Mycenaeans, was an archaic form of Greek. But Linear B is not a literary script that preserves the ideas or myths and legends of the Mycenaeans; rather, it records inventories, lists of people and their belongings, and other bureaucratic matters.

In spite of its literary limitations and the wide debate on Mycenaean influence on the Homeric poems, enough may be gleaned from the tablets to support the idea of a cultural bridge extending from Crete to the Mycenaeans and down to the time of Homer. In his book *The Decipherment of Linear B*, John Chadwick says, 'At present there are two schools of thought: those who believe that the Mycenaean element in Homer is great, and those who think it is small. A compromise here is possibly the best solution. We cannot deny that many features of the Homeric world go back to Mycenaean originals.'[2] For example, among all the inventories on Linear B tablets, there are references to a number of the gods that appear in the *Iliad*. 'The recognizable deities are the familiar names of classical Greece: Zeus and Hera (already coupled), Poseidon, Hermes, Athene, Artemis. *Paiawon* is an early form of *Paian*, later identified with Apollo; *Enualios* is likewise later a title of Ares . . . Aphrodite is so far absent from the texts, but this may be mere chance.'[3] The Linear B tablets list sacrifices made to these gods – an indication that their roles in

Mycenaean society were as important as in the time of Homer and long afterwards.

That Homer had geographical knowledge of Minoan Crete is confirmed in Book 2 of the *Iliad*, when he lists, 'Knossos, and the well-walled city of Gortyn; Lyctus also, Miletus and Lycastus that lies upon the chalk; the populous towns of Phaestus and Rhytium' (2.646). In Book 16 Meriones, second-in-command of the Cretan contingent at Troy, is described by the Trojan Aeneas as a 'good dancer' (16.617); this may be a reference to the athletes who somersaulted over the heads of bulls and whose activities were preserved in frescos on the palace walls at Knossos.

Homer's knowledge of Bronze Age Mycenae, several hundred years before his own era, is found throughout the *Iliad*. The narrative element of the epic tells not of the Iron Age, when the poet-scientist lived, but of an age of heroes, thought to be based upon a romanticized idea of life in Mycenaean Greece. The warriors of the *Iliad*, armed in the fashion of the Bronze Age, were believed in popular memory to have lived at a time when the Mycenaeans ruled the Aegean and influenced events far beyond. In Chapter 9, Homer's knowledge of the geography of Mycenaean times will be seen to be extensive.

In the absence of a script capable of communicating ideas and literature, oral myths provided the means to preserve the astronomical culture of the gifted peoples who for more than a thousand years dominated the Aegean. We believe that Homer was the greatest of the poet-scientists who passed down through the ages the knowledge of the heavens accumulated over many centuries. W. F. Jackson Knight, in his book *Many-Minded Homer*, reminds us that 'Homer comes at the end of a long poetic history. Greek epic verse was hundreds of years old when he used it in the ninth or eighth century BC. It may have been a thousand years old, and have begun its life at about the time, not many generations after 2000 BC, when Greek was first spoken in Greek lands.'[4]

Memory and the Iliad

In a world that has for so long been dominated by the written word and the printing press, it is hard to think oneself back into a past where literature and learning had to be committed to memory so that they could be recalled at will. Memorization was essential to preserve the accumulated knowledge of ancient peoples and to ensure that it was passed on accurately to later generations.

The techniques used to memorize the *Iliad* and the *Odyssey* have long exercised academic minds, and in the 1930s Milman Parry, a young American scholar, made a study of how Homer's use of epithets, formulaic phrases, similes, repetitions and emotive scenes would have guided the ancient poet-singers through their recitations. He corroborated his conclusions in a study of oral poets living in Yugoslavia, where a tradition of reciting lengthy narratives still survived.[5] Since his death in a shooting accident at an early age, Parry's work on the techniques used for preserving stories and legends in memory over long periods of time has been continued by others. And elsewhere in Europe the oral tradition – a worldwide phenomenon – has survived into the twentieth century in societies such as Ireland and, perhaps less surprisingly, Crete. The father of the Cretan writer and Second World War resistance hero George Psychoundakis is said to have been able to recite from memory ten thousand lines of a seventeenth-century epic poem.[6]

To recite the *Iliad*, from the anger of Achilles in the opening lines, to the close when Hector's body is taken back to Troy, takes some twenty-four hours – rather longer with dramatic embellishments. Committing the epic's more than 150,000 words to memory would be a formidable task even with the aid of the features described by Parry. It is hardly surprising that the mother of the Muses in Greek myth was said to be Mnemosyne – Memory – from whose name the word 'mnemonics' is derived for the art of memory training. Frances

Yates, an authority on the subject, wrote, 'One can imagine that some form of the art [of memorization] might have been a very ancient technique used by bards and story-tellers.'[7] It is unlikely ever to be known exactly what techniques the Homeric bards did use to remember not only the *Iliad* but also the more than 115,000 words of the *Odyssey*; however, some insight may be gained from accounts of memory training in post-Homeric Greece.

The poet Simonides of Ceos (*c*. 556–468 BC) is said to have believed that memory could be assisted by assigning striking images to the data to be remembered and then placing them in the rooms of an imaginary house. By letting the mind wander through these 'rooms' when required, the images and the knowledge associated with them could be recalled. Once, when reciting a poem at a banquet, Simonides supposedly used this method to remember where each guest was sitting. After he left, the roof collapsed and crushed the guests beyond recognition, but Simonides identified the bodies by recalling where everyone had been placed at the table. Cicero, the Roman orator, used this story to instruct his pupils that items to be memorized should be placed in order, and the order then be so strongly imprinted on the mind that items could be drawn from their 'place' at any time.

Another memory technique, introduced into Greece by Metrodorus of Scepsis, a contemporary of Cicero (106–43 BC), used an old Egyptian method based not on the rooms of an imaginary house, but on the twelve constellations of the zodiac. If an orator or poet needed to memorize a large amount of material, he could use the twelve zodiacal constellations, subdivided into 360 separate storage places. One constellation could be used to store information on astronomy, another to store information on farming, and another to store myth and legend, and so on.

Memorization takes on an added dimension when Homer's epics are shown to be not just stories but also the repository of a vast amount of astronomical knowledge. Not only did the Homeric bards have to maintain the flow of literary recitation, they also had to understand and present a large volume of learning about the skies. However, Homeric astronomy is so well structured that it too lends itself to being stored in rooms or compartments in the mind. It has five core areas, and the data from each could be stored in a 'house', each house having many 'rooms'. One 'house' could be devoted to Homer's basic catalogue of stars and constellations, with each of its rooms being used to store data about one of the forty-five constellations he describes. A second 'house' for Homeric astronomy might store ideas about the universe, with each of its rooms containing an allegory which when linked to others would create an image of Homer's cosmology. Similar houses could store allegories and learning about the astronomical roles of the gods and about the slow changes in the heavens.

Devices for memorization are useful, but it is the power of Homer's imagery that is crucial in retaining knowledge, and it will be seen in later chapters that the allegories of the *Iliad* are the perfect vehicles for preserving astronomical learning.

It is not a new idea that myths should be examined for a deeper meaning than their surface narrative. From the golden age of Greece, the fifth century BC onwards, the Homeric texts were studied as philosophical allegories, and Alexander Pope wrote in the preface to his translation of the *Iliad* in 1715, 'If we reflect upon those innumerable knowledges, those secrets of nature and physical philosophy which Homer is generally supposed to have wrapped up in his allegories, what a new and ample scene of wonder may this consideration afford us!'[8] W. F. Jackson Knight expands on this:

> There was much talk in ancient times about allegory, whether, and why, the myths and symbols of Homer should be, or should have been, taken to mean something that they did not say. The controversy is very enlightening by its uncertainties. Everything

48

is allegory; and nothing is. Greek science and philosophy grew out of myth and observation combined. The Greek thinkers at first used mythological thought and language to express their conclusions. Myth gave them a scheme to follow, and they used it to develop their rational structures, by substituting abstract conceptions for the concrete personalities in the myths.[9]

The Enigma of Homer

Scholars may argue about the details of Homeric narrative, but they are united in the view that the *Iliad* and the *Odyssey* deserve their place among the world's finest literature. How strange it is, then, that, in spite of his fame and influence, nothing is known about Homer and there is no conclusive evidence that a man of that name ever lived. Some scholars have argued that the *Iliad* and the *Odyssey* were written by the same person, while others have claimed that each epic has a different author or that they are compilations of stories handed down over the ages.

A modern view suggests that Homer – if such a person ever existed – lived *c.* 745–700 BC, and for the purposes of this book these are assumed to be his dates. The present versions of the *Iliad* and the *Odyssey* have their origins in texts collected by Peisistratus, the tyrant of Athens, who died in 527 BC, some two centuries after the poet is thought to have lived. The oldest surviving manuscripts are copies of these texts. Study of Greek dialects indicates that the epics were composed on the west coast of Asia Minor – in modern Turkey. The island of Chios and the town of Smyrna (Izmir) are among a number of places that have claimed Homer as their own.

The literary outline of the Iliad

Just before Paris, son of King Priam of Troy, was born, his mother dreamed that he would be the cause of the city being burned to the ground. He was therefore abandoned to die on a hillside. However, he was found and brought up by shepherds, and was later selected by Zeus to choose who was the fairest of three goddesses, Athene, Hera or Aphrodite. He chose Aphrodite, because she promised him the love of any woman he wished for and also beguiled him with a description of the beautiful Helen, consort of King Menelaus of Sparta in Greece. Paris abducted Helen, and in an attempt to restore her to her husband a Greek army laid siege to Troy – the event that eventually led to the fulfilment of the prophecy. Homer's epic describes the last few weeks of the siege, nearly ten years after the Greek army, led by Menelaus' brother Agamemnon, King of Mycenae, first sailed in more than a thousand ships to the stronghold of the Trojans.

In the first of the twenty-four 'books' or chapters of the *Iliad*, Apollo is laying waste Agamemnon's army in revenge for the King's refusal to return his favourite slave-girl, Chryseis, to her father, a priest of the god. Agamemnon then gives up the girl, but, to save face among his troops, he takes in her place Briseis, the slave-girl of Achilles, the mightiest warrior on the field of battle. Angry at this humiliation, Achilles withdraws from the war to sulk.

After Homer has listed the twenty-nine Greek and sixteen Trojan regiments and their commanders, the fighting erupts again, and the deaths of scores of warriors are described in often gory anatomical detail as the fury of war ebbs and flows.

With the Trojans in the ascendancy, Agamemnon pleads in vain with Achilles to return to the fighting. Achilles refuses, but does let his close friend Patroclus borrow his armour and rejoin the battle. After killing many Trojans, Patroclus is killed by Hector, the foremost Trojan warrior and brother of Paris.

Angered at his friend's death, Achilles is given a new suit of armour and shield by his mother, the goddess Thetis, and he returns to the battlefield to seek revenge. When he comes face to face with Hector, the Trojan at first tries to run away, before standing to fight and meet his fate. Hector is tricked by the goddess Athene and is left unarmed to face Achilles, who dispatches him with a lance thrust to his throat.

In time-honoured tradition, Achilles sponsors funeral games in which such events as a chariot race, wrestling, running and archery are held in memory of Patroclus. The final book of the *Iliad* is the touching journey of King Priam, the father of Hector, to the hut of Achilles to successfully plead for the return of his son's body. The *Iliad* then concludes before the episode of the Trojan Horse and the final Greek assault on Troy.

We have referred to 'Greeks', but the *Iliad* was composed long before Greece was a unified country, and nowhere does Homer use this word. Apart from their tribal or regimental names, he uses the collective names 'Achaeans', 'Argives' and 'Danaans' for the men from the mainland and islands of the region now known as Greece. This book follows popular practice by using 'Greek' in preference to 'Achaean'. The Trojans from the citadel of Troy were supported by their Dardanian relatives and Trojan allies, mainly from Asia Minor.

Principal characters in the Greek forces include:

— King Agamemnon – commander-in-chief of the Greek army;
— King Menelaus – Agamemnon's brother, who lost his wife, Helen, to Paris of Troy;
— Achilles – the most powerful warrior in the two armies;
— Great Aias – the second-most powerful warrior in the two armies;
— Odysseus – a man of influence who comes into his own in the *Odyssey*;
— King Nestor – a wise old warrior who reminisces about the past;
— Diomedes – a strong warrior, much involved in the fighting.

The Trojans include:

— King Priam of Troy – elder statesman and nominal leader of the Trojans (he takes no part in the fighting);
— Hector – the eldest son of King Priam and the strongest warrior in the Trojan army. A very human character, he knows the Trojans are fighting for a lost cause;
— Paris – Hector's brother, a feckless, vain man who stole Helen from Menelaus of Sparta;
— Aeneas – a cousin of Hector who survived the sacking of Troy and, according to legend, founded the city of Rome.

Helen of Troy survived the war and went back to live with her former husband, Menelaus.

The gods who intervene in the action range in character from majestic and honourable to petulant and devious. They are divided in their loyalties between the Greeks and Trojans. Athene and Hera, in particular, have a consuming hatred of the Trojans because Paris had chosen Aphrodite as more beautiful than themselves. The Trojans are favoured by Aphrodite and by Ares, the god of war.

The eventual destruction of Troy is never in doubt after Book 1, when Zeus promises Thetis, mother of Achilles, that her son will have a glorious war, but this bare-bones summary of the *Iliad* gives no sense of the subtle interplay between Homer's characters, and does not address the moral questions that he poses.

Troy

Where was Troy, and was there a Trojan War? The first question is somewhat easier than the second: the site of Troy is three miles from the Dardanelles or Hellespont, at a place now called Hissarlik by the Turks. Troy was thus strategically placed to control trade through the fast-flowing narrows that separate

Europe from the East and connect the Black Sea to the Aegean. Perhaps to enforce the collection of tolls, Troy would have required military power and in turn would have been subject to the envy and resentment of other powers. Other suggested sources for the city's wealth have included a thriving fishing industry and horse breeding, the latter perhaps explaining Homer's epithet of 'tamer of horses' for Hector of Troy (24.804). Such a rich city would have been known throughout the Aegean and beyond, wherever traders discussed their deals and bemoaned whatever tolls may have been levied by the Trojans, and wherever sailors complained about the time they wasted lying at anchor waiting for a favourable wind to help them through the straits.

The first settlement at Troy was founded in about 2500 BC, and nine levels of continuous human habitation have been discovered, the ruins of one level becoming the foundations of the next. Despite more than a century of archaeological research on the site, the detritus of war has not been found in any significant quantity. Precisely which of the many Troys on the site was Homer's is a matter for speculation, but at about 1220 BC one of the levels of habitation, known to archaeologists as Troy VIIa, was destroyed by fire. This has been said to be the real-life event that influenced Homer to use the destruction of a rich and powerful city for his epic.

That Troy burned is without doubt, but who put the city to the torch? There is no convincing evidence that the Siege of Troy so graphically described in the *Iliad* ever happened. Nor is there any proof that Greek and Trojan leaders such as Achilles, Agamemnon, Hector and Paris, or any other characters in the *Iliad*, ever existed except in the imagination.

A more likely explanation for the burning of Troy VIIa may lie with a mysterious force of marauders known as the Sea Peoples, who at that turbulent period in history are thought to have created widespread devastation in mainland Greece, Asia Minor, Cyprus and Syria. They were finally destroyed by the Egyptians in a land battle and sea action in the Nile delta *c*. 1180 BC, but they took with them to the grave all knowledge of their origins.

It is not much more than a hundred years since Troy was considered to have existed only in the realm of myth and legend. Then came a crucial development in Homeric studies when Heinrich Schliemann, a wealthy German entrepreneur, had the notion that the *Iliad* and the *Odyssey* were not entirely stories of fantasy. In 1868 he declared that Troy would be found at the Bronze Age site of Hissarlik, and in succeeding years he carried out dramatic excavations to justify his belief.

Schliemann laid great emphasis on the importance of stratigraphy – the order of different layers of evidence found at a dig – and the importance of pottery finds for establishing a timescale. But he was also a flamboyant showman who wove a web of deception around his life, and probably around some of his discoveries. He has been blamed for destroying important evidence in his haste to drive a huge trench through Troy, and is also said to have falsified records of the discovery of the 'Jewels of Helen' – precious artefacts from an earlier age that he claimed to have found in the ruins of 'Homeric' Troy.

Inspired by his discoveries at Troy, Schliemann reasoned that other cities named by Homer must also exist, and in the following years he excavated in the Peloponnese – at Mycenae (the so-called fortress of Agamemnon) and at Tiryns. It was in a tomb at Mycenae that he discovered a golden mask that is the most famous find from Bronze Age Greece. It was said that Schliemann telegraphed to the King of Greece, 'I have gazed upon the face of Agamemnon.' Archaeologists are now certain it is not the mask of Agamemnon, and there is even doubt about the precise wording of the telegram.

In spite of his many faults and fantasies, Schliemann's were the greatest Greek archaeological discoveries ever made by one person, and the criticism of his early excavations at Troy contains more than a degree of hindsight. When he began work, he was a pioneer of a new science and had no guidelines to follow. In fact he made a major contribution to the creation of archaeology as a discipline, and the sites of Minoan and Mycenaean settlements also written about by Homer were subsequently excavated in Greece, Turkey and Crete.

Archaeological work continues today at Troy, but nothing has been found to indicate that a great Greek army of many thousands of men laid siege for ten long years before razing it to the ground. The following chapters will show that the answer to this mystery of the past two millennia or more lies in the heavens and reflects the importance of astronomy to ancient peoples.

3

Trojan War on Earth and in the Skies

> Read not to contradict and confute; nor to
> believe and take for granted; nor to find talk
> and discourse; but to weigh and consider.
>
> Francis Bacon, 'Of Studies' (1625)

As the *Iliad* progresses, the Greek and Trojan armies battle furiously around the citadel of Troy, and great deeds of heroism are tempered with human frailties, while the gods intervene in the fortunes of both sides. But for two thousand years or more it has been forgotten that Homer had an ulterior motive in describing all this action – to present ancient astronomical knowledge through the medium of unforgettable stories.

In fact, as we shall show, the gods, Greek and Trojan warriors and other characters in Homer's epic represent planets, stars and constellations, and by their activities they preserve a remarkable volume of astronomical learning, encompassing everything that can be seen in the heavens by the naked eye. In purely astronomical terms, Homer's aim was to enable his audiences to identify the stars and constellations and to know their relative positions in the sky. Second, he was aware of significant astronomical events that took place long before his day, and he uses them in chronological order as examples in an exposition of how the skies change over long periods. Reasonable conclusions drawn from observations and inherited knowledge from many previous generations of stargazers in the eastern Mediterranean and Middle East enabled him to illustrate how the heavens slowly changed during a period that began before 8000 BC and ended at about 1800 BC.

Growing familiarity with Homer's techniques for preserving astronomical knowledge encourages more revelations to tumble from every page; the killings of warriors and the intrigues of the gods are revealed as allegories for vivid images in the sky. There is such a breadth and depth to the astronomical content of the *Iliad* that it is almost overwhelming.

Conventional Homeric scholarship and our studies of the *Iliad* are not too different in technique: both interpret layers of narrative to reach conclusions which are not at first apparent. For instance, classical scholars make much of a code of honour, but Homer never codified the conduct demanded from warriors, which can be defined only by interpreting narrative from various parts of the book. So too, layers of narrative can be interpreted to reveal descriptions of the heavens.

Both on earth and in the skies, the *Iliad* is a drama with a structured plot in several acts and a cast of hundreds of players. Fate – the dark inevitability of the doom of Troy on Earth – has its equivalent in the eternal movement of the heavens that is beyond mortal control. Table 1 (*overleaf*) gives a comparison of the two scenarios.

The practical application of astronomy, such as the creation of calendars, can be made only after the stars and constellations have been identified and an understanding of the nature of the universe has been achieved. This was Homer's purpose in the *Iliad*. Homeric astronomy is not a mere assortment of scientific nuggets dropped at random into the narrative of the *Iliad*, but a unified construction that reflects great depth of observation. Each of the many hundreds of pieces of astronomical data so far interpreted from the *Iliad* supports, and is dependent upon, all of the other astronomical

Table 1 *Literal and astronomical narrative in the* Iliad

	On Earth	In the heavens
The stage	The battlefield outside the walls of Troy.	The universe, where the Earth is suspended in space at the centre of a rotating sphere on which are fixed the stars (see fig. 70).
The plot	The scheming and battles that will bring the downfall of Troy and the Trojan allies. The key to the Greek successes is the return to the battlefield of Achilles, the mightiest warrior of all.	The futile Trojan struggle to halt the long-term changes in the sky, including the changing of the pole star, the 'decline' of Ursa Major, and the return of Sirius (the brightest star) to the skies of Greece.
The cast	Regiments of Greek and Trojan warriors, women, mystics and gods.	Mortals and gods represent stars, constellations and planets.
Rules of the game	The battle is conducted under an unwritten code of honour, respected by warriors on each side.	Homeric astronomy is firmly structured, and rules that apply to one celestial event also apply to other similar events – see the Rule of Wounding and the Rule of Magnitude explained later in this chapter.

material in the epic. The *Iliad* is not, however, constructed like a modern textbook, in which basic information is given first before progressively more difficult material follows. Items of astronomical information are interwoven in several continuing themes throughout the epic. Rather like hundreds of jigsaw pieces scattered around the floor, each detail of Homer's astronomy is colourful and interesting in itself, but only when they are assembled can his vision of the heavens be truly appreciated. To make it easier to understand, we have structured this knowledge into five core areas. An overview of how this is intricately woven into the story of the Siege of Troy is given in this chapter before each aspect is explored in more detail later. The core areas are:

1. Homer identifies some 650 stars and places them in forty-five constellations or star patterns. Each mortal warrior and

woman has a 'personal' star, and artefacts such as armour, chariots and spears are also represented by stars linked to these. (Chapters 4 and 5.)

2. Homer acknowledges the ancient supernatural accounts of the creation of the world, but a new generation of gods, the Olympians, led by Zeus, are used as astronomical icons. (Chapter 6.)

3. The precession of the equinoxes – the phenomenon whereby the constellations in which the sun rises at the vernal and autumnal equinoxes slowly change. (Chapter 7.)

4. Homer's concept of the Earth isolated in space and surrounded by a celestial sphere. (Chapter 8.)

5. The use of shapes of constellations as 'skymaps' to guide travellers throughout Greece and Asia Minor. At last it is known how ancient peoples journeyed with confidence and accuracy over long distances. (Chapter 9.)

In addition to presenting these topics, the narrative of the *Iliad* also records celestial phenomena such as meteor showers, solar and lunar eclipses, the retrograde motion of planets, and a likely supernova. Occultations, where an object distant from the Earth, such as a star, is temporarily obscured by one closer to it, such as the Moon or a planet, are also noted.

If the first three points are now considered in more detail, it will be seen how astronomical data and theory are accurately and imaginatively expressed in epic narrative by Homer.

Homer's Catalogue of Stars and Constellations

The great stratagems and ideas of Homeric astronomical theory could only have evolved after form and order had been brought to all the stars visible to the naked eye. The way in which Homer uses five ingenious methods to identify stars and constellations is explained in Chapters 4 and 5; for the moment it is sufficient to state the primary hypothesis that each warrior in the *Iliad* has a 'personal' star in a 'regimental' constellation.

It was Edna Leigh who was first inspired to realize that the 'Catalogue of Ships' in Book 2 of the *Iliad* is not only a roll-call of the forty-five Greek and Trojan regiments but also the foundation of Homer's star and constellation catalogue. Each of the twenty-nine Greek and sixteen Trojan regiments which fought at Troy is identified as a constellation, and the commanders of those units are the brightest stars in their respective constellations. Achilles, for instance, is the greatest warrior at Troy and represents the star known as Sirius, α Canis Majoris, the brightest star. His regiment of Myrmidons is the constellation of Canis Major. From this starting point emerges a catalogue that currently includes some 650 stars in Homer's forty-five constellations. How Homer enumerates and makes memorable the stars and constellations in the *Iliad* and the way in which we gradually unravelled his methods are described in Chapter 4.

Apart from their fundamental importance in the construction of his catalogue of stars, Homer also used some of the constellations for quite a separate purpose. When the poet-singers of ancient times told the stories from the *Iliad* that explained great theories about the movements of the heavens and the place of the Earth in the universe, the night sky was their blackboard. When, for instance, Achilles chases Hector across the heavens to his death (see pages 72–7), it would not have been easy to visualize this event by observing the passage of just two personal stars amid the many hundreds of others that were visible. Homer overcame this problem by letting the characters involved in such important astronomical lessons assume the mantles of their entire constellations when expounding upon such important ideas as the nature of the universe. How much more vivid and instructive it would have been to watch Canis Major (Achilles) pursue majestic Orion (Hector) across the skies! How narrative describes leading warriors in their roles as constellations is examined in Chapter 5.

When we observed that similar events in the narrative of the *Iliad* led to similar astronomical conclusions, we formulated the *Rule of Wounding* and the *Rule of Magnitude*. The Rule of Wounding led to the association of scores of stars with lesser warriors beyond the seventy-three commanders listed in the Catalogue of Ships. Under this rule, a man who is wounded in the eye, for instance, is allocated to a star in the 'eye' of his regimental constellation, and a man struck in the chest is placed as a star in that part of the constellation's anatomy. The effectiveness of this rule is remarkable.

The Rule of Magnitude emerged when it was observed that in all the killings in the *Iliad* the victims were less powerful warriors – or less bright stars – than their attackers. This created a hierarchy in which the brightness of one star could be compared to that of another. When stars representing the killers and their victims are cross-checked, an impressive list of differences in magnitude, some of them quite fine, emerges.

As we shall see, the star catalogue compiled using these rules and derived from the Catalogue of Ships bears striking similarities to the *Almagest*, the listing of stars and constellations by the famous Greek astronomer Ptolemy, who worked in Alexandria in the second century AD and whose writings influenced Western thought until the Renaissance and beyond. Ptolemy, too, identified stars with parts of the 'body' of a constellation, and his model of the Earth at the centre of the universe was similar to that of Homer, though greatly enhanced by the application of mathematics and geometry.

The Catalogue of Ships is said to be the oldest part of the *Iliad*,[1] but it brings a mixed response from readers. For some it is a tedious list of warriors, their regiments and home towns and the number of ships in which they sailed to Troy, offering little to those keen to move on to the clamour of war in the following books. Among classicists, the catalogue has provoked many questions about its origins and purpose. Opinions range from those who suggest that its geographical content is 'a poor invention, interpolated into the *Iliad* by a late and decadent poet, to those who see it as the miraculously preserved record of a historical expeditionary force'.[2] For our purposes however, the catalogue is the single most important part of the *Iliad*, from which flow all other Homeric astronomy and much of the Homeric geography that is explored in Chapter 9. Containing such fundamental knowledge, it should not be surprising that it contains the oldest material that Homer wove into his epic.

Astronomical Roles of the Gods

At the apex of the hierarchy of deities in Homeric astronomy are the supreme god Zeus and his brothers Poseidon and Hades. They rule the visible skies, the seas and the part of the celestial sphere that lies below the horizon. Other gods represent the Moon and the visible planets, as well as the Milky Way and the horizon – the 'river' that was said to encircle the Earth. Zeus' consort, Hera, is the Moon, and four of the five planets visible to the naked eye are represented by his immortal children, the fifth being assigned to Poseidon:

Aphrodite = Venus
Ares = Mars
Athene = Jupiter
Apollo = Mercury (when visible at dawn)
Hermes = Mercury (when visible at dusk)
Poseidon (the 'Earth-girdler') = Saturn

It can be seen that Mercury is assigned to two gods: Apollo, the herald of light, and Hermes, the herald of darkness. This is because Mercury, the planet closest to the sun, can only be seen periodically for a short time either just before dawn or just before dusk. It is not known when it was realized that Mercury was a single planet, but narrative in the *Iliad* indicates that

Homer perceived Mercury's appearance at dawn and dusk as being that of two planets. Venus is another planet that can only be seen before either dawn or dusk and is never in the sky in the middle of the night; narrative indicates that Homer recognized Venus at morning and evening as the same object. In Chapter 6 we give profiles of Hera as the Moon, Athene as Jupiter, Ares as Mars, Aphrodite as Venus, Apollo as Mercury at dawn and Hermes as Mercury at dusk, as well as of the smith-god Hephaestus as the constellation Perseus, and of Thetis, Achilles' goddess mother, as the constellation Eridanus. There is no portrait of Zeus, controller of the heavens and the most powerful god of all, for he does not represent a physical aspect of the universe but is an all-knowing abstract concept. Even his homes on Olympus and Mount Ida are important astronomical ideas rather than tangible entities – see Chapters 6 and 8.

Precession and the Wobbling Earth

The central astronomical theme of the *Iliad* concerns precession of the equinoxes and phenomena that are caused by the slow wobble of the Earth as it journeys through space. It is not suggested that Homer knew the cause of these phenomena, but he was aware of three visible effects that they had on the celestial sphere. To add to his difficulties in presenting these in allegory, he had to contend with the fact that each took place during different historical periods. He overcame the problem by using separate allegories to describe each effect, and wove them into the wider picture of Homeric astronomy. One allegory describes the decline of a north pole star from about 2800 BC to around 1800 BC; another is an exposition of how the equinoxes precessed between the ninth and third millennia; and a third goes back to the ninth millennium, when Sirius, the brightest star, returned to the skies of Greece after a long absence. Following are the three observable effects that Homer preserved in the *Iliad*:

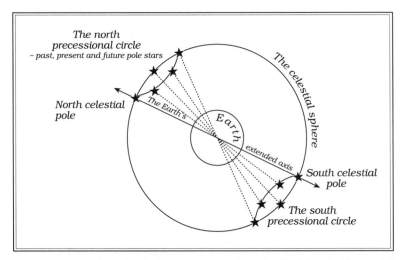

Fig. 6 *In the northern hemisphere, the point where the Earth's axis, if extended, would meet the celestial sphere is called the north celestial pole. The star closest to this point is called the pole star, and will change in time. In the southern hemisphere the Earth's extended axis meets the celestial sphere at the south celestial pole.*

1. The point where an extension of the Earth's axis would meet the celestial sphere is called the celestial pole, and the star nearest to it is called the pole star. This, however, is not a fixed point. Because of the gravitational pull of the Sun and the Moon, the Earth wobbles as it orbits around the Sun, and as a result the imaginary tip of the Earth's extended axis describes a huge 'precessional circle' on the celestial sphere, completing the circle once in about 26,000 years (see fig. 6). As the celestial pole moves around the precessional circle, it comes closer to different stars, so the pole star will change repeatedly during this period (see fig. 62). The pole star in the period of which Homer writes was Thuban in the constellation Draco, but Polaris in Ursa Minor now indicates celestial north.

2. Another effect is that the zodiacal constellations in which the Sun rises at the vernal and autumnal equinoxes slowly change. For example, at the vernal equinox in 2000 BC the

64

Sun rose in Aries, but the vernal equinox at the end of the twentieth century is in Pisces, and in a few more centuries it will be in Aquarius. This effect is known as the precession of the equinoxes, and Homer records the precession of the vernal and autumnal equinoxes over three 'generations', or the precession of each equinox through three constellations. Precession, of course, affects not just the constellation in which the equinoxes lie: all of the twelve divisions of the zodiac 'precess' at the same time.

Precession is a complex matter, and, despite the Greeks' observational skills, it is as certain as can be that Homer would not have been familiar with its modern explanation.

3. Only at the equator can the entire celestial sphere be observed: at other latitudes a part of the celestial sphere cannot be seen – in the northern hemisphere, this will be around the south celestial pole. The size of this region depends upon the latitude of an observer (see fig. 4). However, as precession continues over long periods, stars that once could not be seen at a particular latitude again come into view on the horizon, while others vanish. The Greek astronomer Hipparchus is said to have discovered the precession of the equinoxes in the second century BC, from the appearance of a new star, but evidence from the *Iliad* shows that its effects had been known long before that time. Homer uses the return to the fighting at Troy of Achilles, the *Iliad*'s greatest warrior, as an allegory for the reappearance in the skies of Greece of Sirius, α Canis Majoris, the brightest star.

These three effects of precession were woven by Homer into the three principal strands of the *Iliad*:

1. The fall of the city of Troy is an allegory for the 'fall' of Ursa Major, the constellation that represents Troy. When the pole star Thuban, in the relatively unspectacular constellation of

Draco, was at its closest to celestial north, *c*. 2800 BC, that part of the sky was dominated by Ursa Major. As the celestial pole moved away from Thuban, so too did Ursa Major begin a slow decline from its pre-eminent position adjacent to celestial north. Thuban's tour of duty as pole star ended at about 1800 BC (see figs. 7 and 8).

2. Battles and duels between certain leaders of Greek and Trojan regiments are allegories for the movement of an equinox from one zodiacal constellation to another during a period beginning in the ninth millennium and ending at about 2200 BC.

3. The return of Achilles to the field of battle is an allegory for the appearance in the skies of Greece of the star Sirius, *c*. 8900 BC.

CHANGING OF EQUINOXES AND SOLSTICES

If ancient peoples had not known of the effects of precession, they would have been greatly troubled by observations of the skies over long periods of time. In perhaps as little as three mortal generations it could have been noticed that at the vernal equinox the Sun was not rising at quite the same place within a constellation that it once had. At its furthest – and entirely unlikely – extreme, farmers might have continued for ever to plant their spring crops when the Sun at the vernal equinox rose in Cancer, even though by 1800 BC the Sun was rising in that constellation in June, not March. As the millennia progressed, the sun rose at the vernal equinox in Gemini before moving into Taurus, and by 1800 BC it had moved into Aries. Cancer was then the constellation of the summer solstice – a time far too late for planting.

The twelve constellations of the zodiac represent Greek and Trojan regiments, as explained more fully in Chapter 7. When the regimental leader of a zodiacal constellation is killed, it signifies the movement of either the vernal or the autumnal equinox, or the winter or summer solstice, from one constellation to another. Such is the inevitability of fate, or of precession, that even the supreme god Zeus does not intervene to prevent his mortal son Sarpedon being killed in an allegory that explains the passing of the vernal equinox from Gemini to Taurus.

CHANGING OF A POLE STAR

Over a few generations, it would have been noticeable to ancient peoples – particularly those who sailed the seas or travelled long distances over land – that a pole star, around which the entire heavens appeared to rotate, was not as permanent an indicator of celestial north as might have been thought. This is because, as the centuries pass, the celestial pole slowly moves around the circle of precession away from one 'pole' star and closer to another (see fig. 6). Homer appears to have had an understanding of the precessional circle in the heavens, and of how an imaginary staff piercing the Earth's poles and extending far out into space would indicate celestial north and south. Staffs and enormous spears play prominent parts in the *Iliad* when in the hands of King Agamemnon, Odysseus and Achilles, and they give Homer the opportunity to reaffirm his ideas on the celestial pole and the movement of the heavens.

The inevitable fall of the house of Troy is an allegory for the decline of Thuban in Draco as pole star, from when it was at its most accurate *c.* 2800 BC to its replacement (by a star called Kochab, β Ursae Minoris) *c.* 1800 BC. This also had an adverse effect on Ursa Major, the prominent constellation that represents Troy in the heavens. As the celestial pole moved away from Thuban, Ursa Major began to decline from its pre-eminent position as the constellation closest to the pole star, as shown in figs. 7 and 8.

All of the past, present and future pole stars are 'owned' by the Greeks (see Chapter 7), and both Hector, the Trojan hero, and Agamemnon, the leader of the Greeks, know that the fall of Troy is as inevitable as the decline of Ursa Major. Using

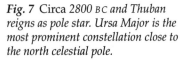

Fig. 7 Circa 2800 BC and Thuban *reigns as pole star. Ursa Major is the most prominent constellation close to the north celestial pole.*

almost the same words in two different parts of the *Iliad*, both the Greek Agamemnon and the Trojan Hector say, 'the day will surely come when mighty Ilium shall be destroyed with Priam and Priam's people' (4.164 and 6.447).

Ursa Major is the Latin name for the Great Bear. Close as the constellation was to celestial north when Thuban reigned, none of its stars can ever be a pole star. Nevertheless its prominence in the northern skies is unchallenged, and '*Arktos*', the Greek name for bear and the name by which the Greeks knew the constellation, gave the region around the North Pole the name of 'Arctic'.

<h2 style="text-align:center">THE RETURN OF SIRIUS</h2>

Homer illustrates a third way in which an effect of precession can be observed: in the reappearance of a star that was once hidden below the horizon at the latitude of Greece. The most spectacular event of this nature happened at about 8900 BC, when Sirius reappeared and was seen by the ancient peoples living in the lands and islands in the region of the Aegean Sea. The arrival of such an outstanding star, after an absence of some seven thousand years, required an equally dramatic

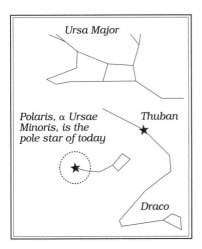

Ursa Major

Polaris, α Ursae
Minoris, is the
pole star of today

Thuban

Draco

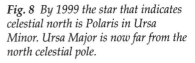

Fig. 8 By 1999 the star that indicates
celestial north is Polaris in Ursa
Minor. Ursa Major is now far from the
north celestial pole.

event in narrative, and for this Homer chose the return of
Achilles to the field of battle. Achilles, the warrior without
equal at Troy, is the literary equivalent of the star without equal
in the heavens.

At the beginning of the *Iliad*, Achilles makes only a brief
appearance before he goes away to sulk after falling out with
King Agamemnon over the possession of a slave-girl. As
Achilles departs, he states, 'They shall look fondly for Achilles
and shall not find him' (1.240). This is no idle warning, for
Achilles is absent from the *Iliad* for much of the fighting, just as
Sirius was absent from the northern skies for countless generations. It then takes two books (chapters) to describe Achilles'
preparations for war and to herald his return to the field of
battle in glittering new armour – an allegory for the return of
brilliant Sirius to the skies of Greece.

Homer's Universe

Homer's cosmology, or model of the universe, was one in which
the Earth was suspended in space at the centre of a star-studded
celestial sphere that daily rotated around it (fig. 9). Whoever first

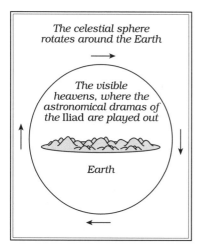

Fig. 9 Homer's concept of the universe, with the Earth at the centre of the celestial sphere.

had the idea of the Earth being isolated in space made what has been called 'the greatest discovery in astronomy'.[3] Homer's expression of the nature of the universe is likely to have evolved by observing how the Sun, Moon, stars and constellations travel in an arc across the heavens. From this it was deduced that, after setting in the west, the heavenly bodies travel in a continuing arc beneath the Earth before rising again in the east. Homer several times describes the passage of stars across the sky, and uses incidents involving both Hector and Achilles to illustrate the passage of constellations under the Earth (see Chapter 8). This is complemented by descriptions of the passage through the zodiacal band of the planets more commonly known as Mercury, Venus, Mars, Jupiter and Saturn (see Chapter 6).

The zodiac is important territory for Homer's gods and mortals: the five planets visible to the naked eye are represented by gods, and the twelve zodiacal constellations are the homes of six Greek and six Trojan regiments. During the time span of the *Iliad*, each of these constellations would have been a marker for an equinox or a solstice. Table 2 shows how narrative extracted from several books of the *Iliad* combines to express Homer's view of the universe.

Table 2 *Narrative and interpretation in the* Iliad

Narrative of the *Iliad*	Astronomical interpretation
1. The duel scenes between Paris (Orion) and Menelaus (Scorpius) at the beginning and end of Book 3 (3.15 and 3.340).	1. Homer delineates half of the celestial sphere visible from Greece with the rising of Scorpius and the setting of Orion.
2. The death of Pandarus (Sagittarius) when he and Aeneas (Virgo) try to kill Diomedes (Perseus) (5.290)	2. Homer completes the sphere by describing the second half, which is fixed by Perseus rising and first Virgo setting, then Sagittarius.
3. Hector goes to look for his wife and to persuade his brother Paris to return to the battlefield (the visible skies) (6.369).	3. Homer 'proves' the sphere by tracing the apparent daily journey around the Earth of Orion, from its setting in the west and going below the horizon to its rising again in the east.
4. Paris says that he will return to the battlefield, but that Hector should 'go first and I will follow and surely overtake you' (6.341).	4. Homer describes how constellations arc across the sky.
5. Zeus tells the lesser gods that, unless they behave, he will 'draw you up with Earth and sea into the bargain, then would I bind the chain about some pinnacle of Olympus and leave you all dangling in mid firmament' (8.23).	5. Homer places the Earth at the centre of the celestial sphere, surrounded by space.
6. Zeus threatens to hurl rebellious gods into the deepest pit under the Earth (the nadir), which is as far beneath Hades as Heaven (the zenith) is high above the Earth (8.16).	6. Homer identifies the zenith and nadir of the celestial sphere – the points directly above and below an observer at any point on Earth.
7. After warning the gods, Zeus leaves Olympus to go to Mount Ida. 'Thereon he lashed his horses and they flew forward nothing loth midway twixt Earth and starry heaven. After a while he reached many-fountained Ida' (8.45).	7. Homer defines celestial north, which from the latitude of Athens is 38° above the horizon – almost midway between the horizon and the zenith.

Some of Homer's projections, such as the setting of the boundaries for the view of the celestial sphere in which most of the fighting takes place, and his vision of the Earth at the centre of the universe, were so important that he did not want them to be lost. To prevent this he created similar images of the heavens by using different allegories. Just as Homer uses the rising of Scorpius (Menelaus) and the setting of Orion (Paris), to establish a view of about half of the celestial sphere, he creates a similar panorama in Book 13 when Meriones, the second in command of the Cretans, 'leaves the field' and fires a bronze-tipped arrow at Harpalion the Paphlagonian, striking him on the right buttock (13.643). This allegory gives a wide panorama of the heavens from θ^1 and θ^2 Tauri (Meriones) setting in the west, as λ Lupi (Harpalion) rises in the east. In Book 11, when Paris fires an arrow at Eurypylus (11.580), the panorama is from the setting of α Orionis (Paris) on the western horizon to the rising of α Lyrae (Eurypylus) on the eastern horizon.

Just as Orion's apparent journey beneath the Earth and across the visible heavens is represented by the movements of Hector, so Homer's description in the opening lines of Book 24 of how the restless Achilles tossed and turned during his sleep can be seen to represent the changing attitude of Achilles' constellation Canis Major as it journeys around the Earth (see fig. 69).

The Death of Hector

The death of Hector is an excellent example of how elements from core areas such as the concept of the universe, the identification of stars and a comparison of their magnitude are brought dramatically together. It reinforces an observation on the way in which constellations arc across the sky by showing that a constellation higher in the sky takes longer to cross the heavens than one closer to the horizon. In this set piece of astronomical narration, Achilles is identified as Canis Major, where Sirius is dominant, while Hector assumes the mantle of Orion.

Hector rejects the pleas of his family and overcomes his own inner doubts and initial cowardice before deciding to stand

and fight Achilles. No one reading of the death of Hector will forget the brutal climax when Achilles spears the Trojan in the throat and gains his revenge for Hector's killing of his friend Patroclus. Nor will they ever look at the night skies in winter in quite the same way again, as Sirius and Canis Major pursue Orion (Hector) across the heavens.

The sequence of events is that (1) Hector stands alone pondering his fate, but (2) as Achilles approaches he decides to run away. (3) Achilles gives chase and (4) 'intercepts' Hector before spearing him in the gullet. The astronomical interpretation of this episode is that the constellation of Orion rises first and is in the sky some two hours before Canis Major (fig. 10). As Canis Major comes over the horizon, it follows on the heels of Orion across the sky (fig. 11). Because Canis Major is lower in the sky, it does not take as long to cross from east to west and 'intercepts' Orion on the western horizon (fig. 12). Orion then sets and vanishes from the sky – or, in literary terms, it 'dies' when Achilles kills Hector (fig. 13).

Achilles' pursuit of Hector across the skies happens each day as the celestial sphere makes one rotation. Canis Major and Orion can be seen in the night skies during the autumn and winter months, but at other times they are in the sky during the day and cannot be seen.

> King Priam was first to note [Achilles] as he scoured the plain, all radiant as the star which men call Orion's Hound, and whose beams blaze forth in time of harvest more brilliantly than those of any other that shines by night . . . even so did Achilles' armour gleam on his breast as he sped onwards. (22.25)

King Priam, from his position high on the walls of Troy (Ursa Major), is the first to see Achilles (Canis Major) approach. Orion (Hector) has only recently risen and is still low in the sky, whereas from Ursa Major Priam would have been able to see Canis Major earlier. Homer is for once being direct rather than allegorical when Priam announces the approach of Achilles, as the 'dog star' Sirius. This is the only occasion in the *Iliad* in which Homer specifically identifies a warrior with his personal star.

> Hector stood his ground awaiting huge Achilles as he drew nearer towards him . . . [Hector] leaned his shield against a tower that jutted out from the wall and stood where he was, undaunted. (22.96)

In this scene Hector/Orion stands alone awaiting the rise of Achilles/Canis Major (fig. 10).

> Thus did [Hector] stand and ponder, but Achilles came up to him as it were Ares himself, plumed lord of battle. From his right shoulder he brandished his terrible spear of Pelian ash, and the bronze gleamed around him like flashing fire or the rays of the rising sun. Fear fell upon Hector as he beheld him, and he dared not stay longer where he was but fled in dismay from before the gates, while Achilles darted after him . . . As a mountain falcon . . . even so did Achilles make straight for Hector with all his might. (22.131)

In the course of a night, this scene takes several hours as Canis Major pursues Orion across the sky (fig. 11) towards the western horizon where they both will set.

> Achilles was still in full pursuit of Hector . . . Whenever he [Hector] made a set to get near the Dardanian gates and under the walls, that his people might help him by showering down weapons from above, Achilles would gain on him and head him back towards the plain, keeping himself always on the city side. (22.188)

As the chase nears its end, or as Orion is about to set, Achilles intercepts Hector. Orion rises some two hours before Canis Major, but Canis Major is lower in the sky, and as Orion (Hector) is about to set, Canis Major (Achilles) cuts off its escape (fig. 12).

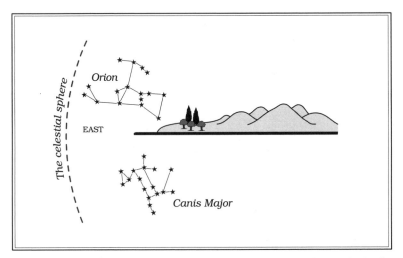

Fig. 10 Hector awaits the arrival of Achilles. Orion (Hector) can be seen in the sky, but Canis Major (Achilles) has not yet risen.

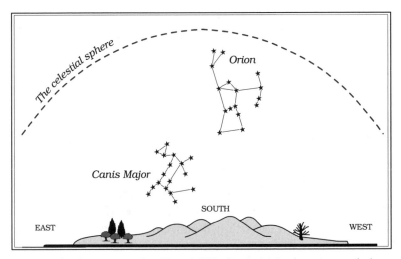

Fig. 11 The chase across the skies. Achilles/Canis Major has risen and chases Hector/Orion towards the horizon in the west.

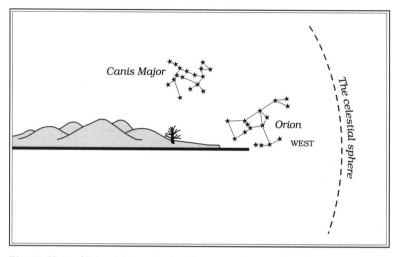

Fig. 12 *Hector/Orion is intercepted on the western horizon by Achilles/Canis Major as both constellations are near to setting.*

[Achilles] eyed [Hector's] fair flesh . . . but all was protected by the goodly armour . . . save only the throat where the collarbones divide the neck from the shoulders, and this is a most deadly place: here then did Achilles strike him as he was coming on towards him, and the point of his spear went right through the fleshy part of the neck. (22.321)

Hector/Orion is dipping below the horizon as a lance flashes across the sky from Sirius, Achilles' personal star, and strikes a fatal blow in the gullet of Orion (fig. 13). The lance thrust to Hector's neck identifies stars in Orion through the Rule of Wounding (page 89); and the principle of the Rule of Magnitude, which says that a star can only be 'killed' by a brighter star, is adhered to in the killing of Hector by Achilles, whose personal star of Sirius far outshines any of the stars in Orion.

Just before death takes him, Hector pleads for his body not to be thrown to the dogs, but Achilles insists 'nothing shall save you from the dogs . . . though they bring ten or

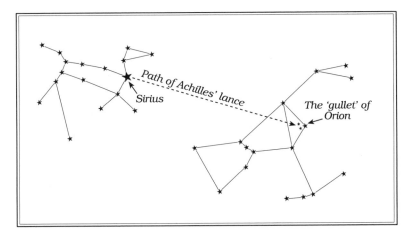

Fig. 13 *Achilles' lance hits Hector in the gullet. When Achilles' lance strikes Hector, it identifies three stars in Orion – the two collarbones and the gullet or throat.*

twenty-fold ransom' (22.348). In a sense, Hector has already been brought low by two dogs; by killing Patroclus (Canis Minor) he brought Achilles (Canis Major) back to the battlefield to gain revenge for his friend's death.

4

Warriors as Stars

As when the stars shine clear, and the Moon
is bright – there is not a breath of air, not a
peak nor glade nor jutting headland but it
stands out in the ineffable radiance that
breaks from the serene of Heaven; the stars
can all of them be told . . . A thousand camp
fires gleamed upon the plain, and in the
glow of each there sat fifty men.

Iliad 8.555

A s this memorable quotation implies, to the ancients the
spectacle of the night skies was as much a part of their
lives as the Sun, the rain and the wind; but there could be little
understanding of the heavens until order was imposed upon
the stars. Constellations, and individual stars within their
boundaries, had to be identified, and with this accomplished
the exercise was completed by recording the relative positions
of the constellations in the sky, so making an atlas of the
heavens. This chapter explores how the people who lived in
the Aegean for a thousand years or more before Homer preserved a considerable volume of astronomical information of
this kind – information that in turn was embedded by the poet
in his epic.

By the late 1950s Edna Leigh was pursuing the idea of an

astronomical context to Homer's epics and had come to believe that the regimental homelands of Greeks and Trojans so carefully described in the Catalogue of Ships in Book 2 of the *Iliad* were in some way connected to constellations in the sky. The Catalogue of Ships is a curious interlude in the narrative, falling between Achilles' dispute with Agamemnon and the resumption of hostilities between the Greeks and Trojans. The lengthy list of the forty-five regiments of Greeks and Trojans who fought at Troy is easily skipped over as the reader heads towards the action in succeeding books. Edna's view, however, was that every word in the *Iliad* had a bearing on the epic's concealed content, and this belief bore a bountiful harvest.

We think the key that opened the door to all her following successes was found when her eye was drawn to the region of the Bay of Pegasus, a large bay on the eastern side of mainland Greece to the north of the island of Euboea, while studying a map. She took the bold but simple step of assuming that the bay, together with an ancient dried-out lake and a pattern of towns to the north, was associated with the outline of the constellation of Pegasus. It was apparent that the position of the brightest stars in Pegasus bore a reasonable similarity to the sites of the named towns on a terrestrial map (see figs. 14 and 15). 'This northern region was agricultural, and Pegasus appears to designate river valleys, plains, and mountain passes between its major stars. Homer frequently mentions Pherae and Mount Pelion, represented by the leading stars in the constellation, each of equal comparative brightness,' she wrote. Superimposing the outline of the constellation on to the map identifies the modern towns of Volos and Velestinon, known in ancient times as Iolcus and Pherae. The large ancient Lake Boebe, modern Lake Karla, is now drained, but it fell naturally into the Square of Pegasus. Other aspects of the constellation directed her eyes to the coastline, to Mount Ossa and to Mount Pelion, while its strangely shaped 'head' and 'legs' projected an image of the Bay of Pegasus itself.

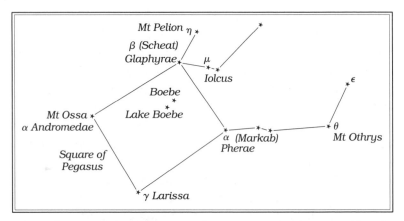

Fig. 14 *The constellation of Pegasus, with stars representing the towns of Eumelus and his warriors.*

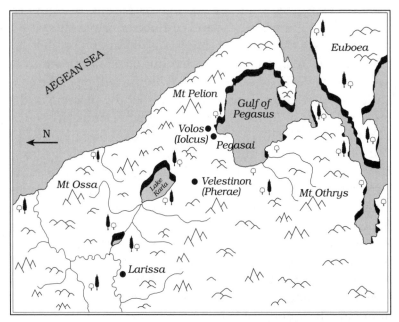

Fig. 15 *The region around the Gulf of Pegasus. The town of Volos is Homer's Iolcus, and Velestinon is Pherae. Modern Lake Karla is the remains of the once larger Lake Boebe, which is now dried up. The road pattern of this region can also be linked to the shape of Pegasus.*

She then turned to the Catalogue of Ships. Of the region surrounding the Bay of Pegasus, Homer says the regiment of men came from 'Pherae by the Boebean lake, with Boebe, Glaphyrae, Pherae and the populous city of Iolcus' (2.711), and they were led by Eumelus. She delved into the *Iliad* for other references to Eumelus, and discovered that the descriptive narrative she found could be linked to the visual qualities of Pegasus and its relationship to adjacent constellations.

Eumelus is famous for his 'matched' pair of horses, described thus in Alexander Pope's translation:

> Eumelus' mares were foremost in the chase . . .
> Fierce in the fight their nostrils breathed a flame,
> Their height, their colour, and their age the same. (2.763)

The E. V. Rieu translation says that the horses were joined by a yoke that was as true as a plumb line. A modern chart of the constellation shows that two of the brightest stars in Pegasus – Scheat, β Pegasi, and Markab, α Pegasi – are indeed the same colour and have the same degree of brightness. What is more, when the constellation is at its highest point as it crosses the sky Scheat lies almost directly north of Markab in the Square of Pegasus and an imaginary line drawn between them and continued down to Earth would make an angle of almost 90° with the horizon. 'True as a plumb line' is thus an excellent literary description for the relationship between these two stars.

Later we were to find further evidence of the astronomical association of Eumelus, when, during the funeral games in Book 23, his yoke 'breaks' (23.392). Scheat, β Pegasi, at one end of the yoke, is a variable star whose fluctuations can be seen with the naked eye. Homer draws attention to this star by the 'snapping' yoke. This astronomical cameo of Pegasus is further embellished when Eumelus then tumbles from his chariot, receives a bloody nose (a star in Pegasus' nostril), and falls beneath the wheel of his chariot – the Circlet of Pisces that lies below Pegasus in the sky.

A map of Greece may have been a roundabout way for Edna Leigh to come to Homeric astronomy, but she had at last shown that there was a dimension to the *Iliad* that had not been suspected for many centuries. After finding an example of how the shape of a constellation could be related to a land area of Greece and associated with characters and incidents from the *Iliad*, she began to look for additional astronomical information from the Catalogue of Ships. She first examined the more famous regiments in search of a connection between them and constellations. King Agamemnon's home in Mycenae, the golden city with its famous Lion Gate, was linked to the familiar lion-like shape of the constellation Leo; the steeply rising Peloponnese mountains along the Bay of Corinth became the Lion's broad back, and an ancient route from Mycenae via Cleonae to Corinth became his front leg (see figs. 76 and 77). The peninsula of Messini, the land of Nestor, the Gerenian charioteer, was associated with five-sided Auriga, the constellation of the charioteer (see figs. 82 and 83). The narrow valley and headwaters of the Eurotas valley (Lacedaemon), the domain of King Menelaus of Sparta, became the spindly constellation of Scorpius (see figs. 74 and 75). The small island of Salamis, home of Great Aias, physically the largest warrior at Troy, was linked to Argo Navis, the largest constellation in Homer's day, but since divided into three (see figs. 72 and 73).

Once the validity of linking geographical land areas to constellations had been established, the way was open for long-forgotten knowledge to be garnered from the remaining regiments. Chapter 9, 'Homer the Map-Maker', investigates more deeply these and other examples of how constellations can indicate routes and topographical details of the lands and islands of the Aegean and Ionian seas.

These studies made Edna aware that there were many levels of learning preserved in the *Iliad*. She had assumed that the shapes of constellations known in ancient Mesopotamia were also familiar in Homeric Greece. She then appears to have reasoned that, this being so, Homer might also have had a method of identifying individual stars. She concluded that, if each of the forty-five Greek and Trojan regiments listed in the Catalogue of Ships represented a constellation, it was probable that the brighter stars in each constellation would represent the named commanders of the troops. The realization that Homer used stars and constellations for different astronomical purposes was to prove of great importance for later interpretation of the epic.

Eventually, with further descriptive evidence gained from narrative of the *Iliad*, Agamemnon, 'King of Men', became the star Regulus from the constellation of Leo; smooth-talking Nestor was associated with Capella in Auriga; red-haired Menelaus represented Antares in Scorpius; Diomedes, of the loud war cry, became Mirphak from Perseus, while Great Aias, the second strongest warrior at Troy, became Canopus (α Carinae) from Argo Navis, the second brightest star in the heavens. Sirius in Canis Major, the brightest star of all, was allocated to Achilles, the warrior without equal on the battlefield. Edna's initial associations had been consolidated by further scrutiny of narrative that showed that the physical attributes given by Homer to some of these leaders matched prominent characteristics of their 'personal' stars. Examples of how phrases from narrative help to build astronomical images of warriors are shown later in this chapter and in Chapter 5.

Among Edna's papers are many pages of notes which reflect her extensive research into myth and the families and ancestry of great heroes in support of her hypothesis. Eventually she had constructed a basic catalogue of stars and constellations from the Catalogue of Ships, but with many other matters of Homeric astronomy pressing for her attention she put aside this work and turned to other topics that had evolved from her primary studies.

Edna once wrote that if she had realized what the outcome of her study was going to be, she would have gone about it in a very different way. No doubt if she had known at the beginning that the geographical content of the *Iliad* was dependent upon knowledge of all the stars and constellations, she would have started by creating a Homeric catalogue of them from the Catalogue of Ships. No doubt, too, she would have avoided the blind alleys and fruitless lines of inquiry that consumed so much time and energy. With the advantage of hindsight, when we took on the task of presenting her work it was decided not to follow the sequence in which her discoveries were made, but to take a more formal approach and begin with the fundamentals of astronomy, the identification of stars and constellations.

Edna had determined that forty-five constellations and seventy-three of the brightest stars in the sky could be identified from the Catalogue of Ships. It became apparent to us after her death that, if commanders had 'personal' stars, then logically every other warrior in the *Iliad* should also represent a star in his 'regimental' constellation. This eventually proved to be so, and we found too that the weapons and artefacts of warriors, together with their wives and slaves, could be identified with stars. Edna's original catalogue of around seventy stars has now been expanded to some 650 stars in forty-five constellations. We say that Homer identifies 'some' or 'about' 650 stars because the exact number cannot be determined. As we shall demonstrate , Homer identifies some stars in groups rather than individually – as the 'armour' of Odysseus or the 'cloak' of Paris, for example – but exactly which stars he saw as being in these groups is not known.

With hundreds of stars identified, another important issue was raised. If stars were named after warriors, what names were given to individual constellations? The answer was again found in the narrative, which showed that some leaders – including Agamemnon, Achilles, Hector, Paris, Diomedes,

Odysseus and Menelaus – had dual roles. Not only did each of them have a personal star, but in some of the best-known scenes from the *Iliad* they could also represent an entire constellation. How this became apparent is examined in Chapter 5.

Homer's division of the heavens into forty-five constellations is similar in number to the forty-five of Aratus in the third century BC and the forty-eight listed by Ptolemy in the second century AD. Since those times, parts of the heavens have been reclassified and new constellations have been created. For instance, faint stars from a part of the constellation known to Homer as Leo were used by Hevelius (AD 1611–87) to create Leo Minor. The nomenclature used in this book is the modern one.

Modern astronomers with powerful telescopes talk of the millions of stars in the universe, but only about 2,000 individual stars down to magnitude 6 can be seen by the naked eye, and even then only under good conditions. Caution has been the watchword in compiling the Homeric catalogue, but the current list of stars may be expanded if future research is able to associate the number 1,206 – the number of Greek ships that sailed to Troy – with individual stars.

The most comprehensive ancient catalogue of stars until now has been Ptolemy's *Almagest* (*c.* 140 AD), which lists 1,022 stars and forty-eight constellations. The *Almagest* influenced Western astronomy until the sixteenth century, and in its method of identifying individual stars it has similarities to the much older catalogue of Homer. Both astronomers placed many stars within constellations according to bodily positions; Homer, for instance, identifies stars as being in the eye, ear, neck or liver of constellations, and Ptolemy used a similar method.

The structure of the *Iliad* both as literary text and as astronomical treatise shows that Homer was well aware of techniques that make it easier to recall large amounts of narrative and data. To overcome the daunting challenge of memorizing hundreds of stars in a single list, Homer broke the stars down into five divisions:

— the brightest stars in a constellation – the commanders of regiments;
— stars linked to second-rank warriors;
— stars placed where warriors are wounded;
— less significant stars represented by warriors who die in clusters or have unobstructed views of their killers;
— stars linked to artefacts such as armour, clothing, shields, chariots etc.

High-Ranking Commanders – the Catalogue of Ships

• *The warrior leaders listed in the Catalogue of Ships are identified as individual stars, and their regiments are the constellations containing those stars. The 'personal' stars of commanders such as Agamemnon, Odysseus and Hector are the brightest in their constellations. For sound astronomical reasons (see Chapter 5), these commanders can also represent their entire constellations.*

The *Iliad* took on a new weight of meaning in 1873, when Heinrich Schliemann uncovered the ruins of Troy. No longer was it considered to be only a fictional story of gods and great heroes; it now had a physical substance. Schliemann and later archaeologists went on to uncover other towns named in Homer's epics whose sites had been long forgotten.

Schliemann's discoveries and claims about Troy stimulated the continuing arguments about the significance of the Catalogue of Ships. This ostensibly simple listing of some of the major players in the *Iliad*, the places where they lived and the number of ships in which they sailed to Troy has generated a great deal of scholarly discussion: about its origins and its relationship to the rest of the *Iliad*, the order in which regiments are presented, and whether all or just part of it was composed by Homer. Nevertheless, arguments have not been very conclusive during the past century or more. In 1915 W. Leaf suggested that the catalogue owed its origins to someone whose aim was to give a place in the Trojan war to the smaller communities and to individuals less involved in the fighting. Another idea has been that Homer's order of listing the regiments was changed by a post-Homeric poet from Boeotia so that soldiers from that region could be given the honour of leading the assault on the beaches of Troy, to give them a significance they do not achieve in later battle scenes. In 1873, E. A. Freeman, who was to become Professor of History at Oxford, wrote, 'We have never doubted for a moment that the Catalogue in the *Iliad* is a real picture of the Greek Geography of the time . . . no conceivable motive can be thought of for its invention at any later time.'[1] Edna Leigh's work on the catalogue confirmed the accuracy of the first part of his statement, but took matters much further.

The catalogue, or 'Display of Forces' as it is sometimes known, lists the officer class at Troy – the commanders of Greek and Trojan regiments. It identifies:

— twenty-nine Greek regiments and sixteen Trojan regiments;
— seventy-three commanders of regiments;
— in many cases, the towns from which soldiers were recruited;
— the numbers of men and ships with each Greek regiment.

When the Catalogue of Ships is viewed as a catalogue of stars, it can be seen that:

— The Greek and Trojan regiments represent Homer's forty-five constellations.

— The commanders and leaders are the seventy-three brightest stars in the constellations. A few examples:

Aeneas = Spica, α Virginis (Virgo)
Agamemnon = Regulus, α Leonis (Leo)
Arcesilaus = Thuban, α Draconis (Draco)
Idomeneus = Aldebaran, α Tauri (Taurus)
Menelaus = Antares, α Scorpii (Scorpius)
Odysseus = Arcturus, α Boötis (Boötes)
Sarpedon = Pollux, β Geminorum (Gemini)

— The shapes of most constellations identify land areas of Greece and Asia Minor.

Edna Leigh's initial work to associate the Greek and Trojan regiments with constellations required extensive knowledge of Homeric epic, myth, astronomy and geography, and progress was painstaking and slow. There were particular problems with some Trojan regiments for whom Homer gives little information apart from the geographical area from where they came. These areas are often little known and are usually on the outer boundaries of Troy's sphere of influence. Examination of other parts of the epic has, however, provided enough information to propose a logical place in the heavens for each one. At the end of Chapter 5 is a summary of the regiments, the constellations they represent and the personal stars of their commanders.

Senior Warriors of the Second Rank

• *Senior warriors of the second rank generally represent the next brightest stars after those of the commanders.*

Many high-ranking warriors play important roles in the narrative of the *Iliad* but are not named in the Catalogue of Ships.

These include the following, who can be identified with less bright stars of the constellation to which their regiment belongs:

Agenor, son of the Trojan elder Antenor = Arich, γ Virginis
Antilochus, son of Nestor = El Nath, β Tauri
Euryalus, second commander from Argos = γ Persei
Paris, son of Priam and brother of Hector = Betelgeuse, α Orionis
Patroclus, comrade of Achilles = Procyon, α Canis Minoris

The Rule of Wounding

• *Each warrior identified by the Rule of Wounding is first allocated to a constellation according to the regiment to which he belongs, and is then placed within that constellation according to where he is wounded – a man wounded in the eye will be in the 'eye' of the constellation.*

Following Edna Leigh's death in 1991, we realized that her work on star and constellation identification was far from complete and that more astronomical data was waiting to be extracted from the *Iliad*.

If the commanders of regiments were the brightest stars, it followed that all the other warriors named in the *Iliad* also had a place in the skies. The remaining men far outnumber the commanders, and many of these warriors die terrible deaths which Homer describes in gory detail: decapitated heads roll across the ground, the bowels of warriors spill out on to the earth, eyes pop out of smashed skulls, arms and legs are chopped off, men are speared in the buttocks, flying arrows pierce the soles of warrior's feet, and even more intimate areas bear the pain and anguish of war. These killings begin in Book 4, after Pandarus breaks the truce by firing an arrow at Menelaus, and they end only with the brutal death of Hector in Book 22.

Florence Wood pondered this for months, and many avenues were explored in an attempt to discover the astronomical purpose of the warrior victims. Men were picked out at random, and every reference made to them was examined for clues which would place them in the heavens. Lists of victims and their killers, of the types of weapons used and of family relationships were created and studied, but a solution remained elusive.

Eventually, late one evening after a fruitless day cross-checking more lists, Florence casually thumbed through an illustrated guide to astronomy[2] and found the answer to the problem in a surprisingly simple drawing. The constellation of Piscis Austrinus (the Southern Fish), shown in the guide, is not one of the more memorable groups of stars, but she had earlier speculated that the brightest star in the constellation, Fomalhaut, α Piscis Austrini, might be identifiable with the warrior Iphition, on the meagre grounds that Homer says his home 'is on the Gygaean lake where your father's estate lies, by Hyllus, rich in fish, and the eddying waters of Hermus' (20.390). But the tenuous link between fish and water had not been supported by other evidence, and the idea had been discarded. A second look at the drawing of the Southern Fish changed everything. It showed the stars in the outline of a fish against the blue background of the heavens (fig. 16). Fomalhaut, with a magnitude of 1.2, was placed exactly on the forehead of the drawing. A few lines earlier in the *Iliad* Homer recounts that 'Achilles struck [Iphition] full on the head as he was coming on towards him, and split it clean in two' (20.387).

At long last we came to realize that Homer's constellations had human attributes such as heads, foreheads, legs, arms, backs and chests. This was the breakthrough in understanding

Fig. 16 *The constellation of the Southern Fish superimposed on to a sea bream, a fish common in the seas of Greece. There is a painting of a sea bream on a fifth-century-BC Greek dish in Manchester Museum.*

which led to the creation of the Rule of Wounding, according to which the place on the body where a warrior is wounded identifies the position of his star in a constellation. The practical application of this rule to all the killings and woundings of the *Iliad* means that, for instance, the personal star of a man wounded in the back is placed in the 'back' of his constellation, and the personal star of a warrior hit in the foot is placed in the 'foot' of a constellation, and so on.

Before a warrior could be placed in a constellation according to where he was fatally injured, it had to be established to which regiment (constellation) he belonged, and this was not always obvious from the narrative. Some warriors who did not belong to an obvious geographical area could nevertheless be organized into logical groups, such as the sons of citizens of Troy, the sons of the Trojan elder Antenor, the sons of the Trojan elder Panthous, and the sons of King Priam. Identifying each of these groups with specific constellations was exacting and time-consuming, but eventually successful.

Of these groups, the sons of King Priam numbered fifty and were allocated to stars in the large constellation of Orion. His legitimate sons are brighter stars in the familiar outline of

Orion (fig. 1), while the bastard sons are designated as stars outside the figure of Orion but within the constellation boundary. Arbitrary as this may at first seem, it proves to have substance when Hector or Paris assumes the mantle of the constellation of Orion to enact a role in Homer's exposition of the precession of the equinoxes and the nature of the universe. So specific is Homer in defining the parts played by the two leading sons of Priam, that, as will be seen in Chapters 7 and 8, their actions could not possibly be allocated to other stars or constellations. The other sons of Priam who play a role in the *Iliad* are placed in Orion according to their wounds. Not all of Priam's sons take an active role in the *Iliad*, but three stars placed in the constellation according to the Rule of Wounding are Helenus, χ^1 Orionis, who is wounded in the hand (13.597), Polydorus, ϵ Orionis, killed by a spear through his belt (20.414), and Lycaon, ϕ^1–ϕ^2 Orionis, who is killed by a blow to the collarbone (21.117).

The constellation of Ursa Major, the citadel of Troy in Homer's explanation of the changing of the pole stars, was a natural home for those wounded warriors described as the 'sons of' citizens of Troy. Using a similar logic, the sons of Antenor were allocated to his constellation of Virgo, and the sons of Panthous given to Libra.

WARRIORS IN URSA MAJOR

To explain how the Rule of Wounding works in greater detail, we will examine how an unusual group of warriors is placed in Ursa Major. Many of the Trojan allies fit easily into their regimental constellations, because Homer tells us specifically that they are Lycians, Mysians, Paeonians, Paphlagonians, Thracians, and so on. The Dardanians and 'sons of' other Trojan elders, as well as those of King Priam, also have clearly defined places in the heavens. Eventually one group of men stood out as apparently without a celestial home, and they all shared a common epithet. Each named by Homer as the 'son

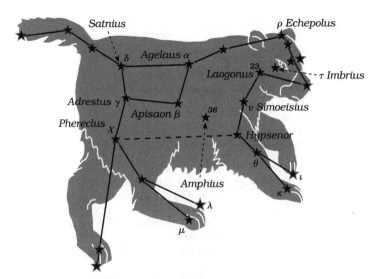

Fig. 17 *'Sons of Troy' located in Ursa Major according to their wounds. (The remaining stars are allocated according to other Homeric methods.).*

of . . .', they included Echepolus, *son of Thalysias*, Simoeisius, *son of Anthemium*, and Hypsenor, *son of Dolopion*, but, unlike other 'sons of . . .', none of them was associated with a particular regiment.

Following the logic that all warriors must have a place in a constellation, an examination was made to see whether they would fit in the only remaining constellation not to have warriors assigned to it. This was Ursa Major, the constellation that, representing the citadel of Troy, dominates the narrative of the *Iliad*. A list was made of all of the unplaced warriors described as a 'son of . . .' together with the location of their wounds. It was soon discovered that there were stars in the correct places of the body of Ursa Major to match the wounds suffered by these 'sons of . . .'.

In fig. 17 the stars of Ursa Major within the body of the bear are identified with specific 'sons of . . .' according to where they were wounded. Table 3 gives details of the sons' injuries.

Table 3 Warriors in Ursa Major

Modern name	Mag.	Homeric name	Cause of death	Reference
ρ	4.8	Echepolus, *son of Thalysias*	Speared by Antilochus in the forehead; the first Trojan killed in the *Iliad*	4.458
υ	3.8	Simoeisius, *son of Anthemion*	Speared by Great Aias in the right breast, near the nipple	4.473
χ	3.7	Phereclus, *son of Tecton*	Speared by Meriones low in the right buttock	5.59
θ	3.0	Hypsenor, *son of Dolopion*	Struck in the arm with a sword by Eurypylus and his hand severed	5.76
36	4.6	Amphius, *son of Selagus*	Speared by Great Aias in the lower belly	5.612
γ	2.4	Adrestus, *son of a rich father*	Captured by Menelaus, but speared in the flank by Agamemnon	6.61
α	1.8	Agelaus, *son of Phradmon*	Speared by Diomedes through the back	8.257
β	2.4	Apisaon, *son of Phausias*	Speared by Eurypylus in the midriff, under the liver	11.578
τ	4.7	Imbrius, *son of Mentor*	Lanced below the ear by Teucer, before Little Aias cut off his head at the neck with a sword	13.170
δ	3.3	Satnius, *son of Enops*	Speared in the flank by Little Aias	14.442
23	3.6	Laogonus, *son of Onetor*	Struck under the jaw and ear by Meriones	16.605

In Homeric astronomy there is a subtle interplay between the Rule of Wounding, the Rule of Magnitude and the location of constellations in the night sky. After the 'sons of . . .' had been placed according to the Rule of Wounding, there were still a number of stars without warriors and a number of warriors without a home. Again these men were affiliated to no regiment, but all could be given a home in Ursa Major that would put them on an unobstructed line of sight to their killers. That left three bright stars in the 'tail' of the Bear, which is also the 'handle' of the Big Dipper. None of the *Iliad's* warriors receives fatal wounds in his 'tail', and so none could be placed in the 'tail' of Ursa Major. These stars were allocated by other Homeric means.

The Catalogues of Homer and Ptolemy

The exhilaration at working out the implications of the Rule of Wounding lasted for eight months, and was enhanced in the summer of 1996 when researching for a few weeks in the libraries of the University of Texas, at Austin. There Florence studied a copy of Ptolemy's *Almagest* for the first time. Sharing a table with students in the calm of the Astronomy Library, it was difficult to contain the great surge of excitement that she experienced when she realized that Ptolemy, too, gave human attributes to constellations and placed many stars accordingly. One of the first warriors placed using the Rule of Wounding was the Trojan Echepolus, who is fatally wounded in the skull (4.458). This placed him at ρ in the head of Ursa Major. Ptolemy's description in Latin of that same star in Ursa Major is 'Praecedens earum quae in fronte sunt' – 'the leading one of those which are in the forehead'.[3] Echepolus' selection for the dubious honour of being the first to die is Homer's way of drawing attention to how ρ Ursae Majoris was the first, or leading, star that preceded the entire constellation of Ursa Major around Thuban, α Draconis, the pole star of the *Iliad*. It might be argued that it is a remarkable coincidence that Homer and Ptolemy used almost identical systems to place stars, but we feel it was another indication of a thread of astronomical learning stretching from long before Homer down to the Greek philosophers of later centuries who applied mathematics and geometry to turn observational astronomy into a more systematic, but often flawed, study.

Ptolemy uses the technique of identifying some stars according to how they were seen in the 'body' of a constellation, but he does not use other Homeric techniques of identification. A reason for this might well be that by the time of Ptolemy the preservation of knowledge in written form was so common that there was no need to divide the stars into different groups in order to make them easier to memorize.

Long before a connection had been made between Homeric

astronomy and Ptolemy, the authors had made an arbitrary decision that Homer's constellations represented warriors seen face to face by observers on Earth. This notion worked well except in a small number of cases where narrative suggested that Homer also described warriors in profile. For instance, Diomedes as Perseus is seen not only face to face but also from the side, or in profile, running across the night sky to the western horizon. Similarly, the ancient circumpolar constellation of Lyra – the Lyre or Harp – is represented by fleet-footed Eurypylus and visualized in profile. Although the identification of constellations both on a face-to-face basis and in profile lent itself to the narrative of the *Iliad* and the logic of Homeric astronomy, we had lingering doubts about its validity, until G. J. Toomer in his translation of the *Almagest* drew our attention to the fact that both Ptolemy, and Hipparchus before him,[4] had seen constellations both facing the Earth and in profile:

> On the matter of the orientation of the figures, I have satisfied myself that Ptolemy describes them as if they were drawn on the *inside* of a globe, as seen by the observer at the centre of that globe, and facing towards him. This is in agreement with what Hipparchus says . . . for all the stars are described in constellations from our point of view, and as if they were facing us, except for such as those which were drawn in profile.[5]

In the catalogues of Homer and Ptolemy, some constellations are more closely comparable than others, and this may be due to the ways in which constellations were perceived in different ages. Astronomers since Ptolemy have rearranged the outlines and boundaries of constellations and from time to time have moved stars from one to another; there is no reason to think that in the nine hundred years or so between Homer and Ptolemy similar changes were not made and constellations were not modified and redrawn. Ptolemy had benefited from several centuries of maturing scientific thought and the transition of astronomy from a science based on observation to one in which mathematics and geometry were applied.

Table 4 (*overleaf*) shows some striking examples of how data from Homer's catalogue is matched by that from the *Almagest* (as described in G. J. Toomer's translation). Ptolemy, like Homer, viewed constellations in configurations also known in Mesopotamia, and these are divided into four distinct types: human form, part man and part beast, animals, and inanimate objects.

In Ptolemy's original list there are in fact nine constellations named after inanimate objects: Crater, Corona Borealis (the Northern Crown), Corona Australis (the Southern Crown), Sagitta (the Arrow), Triangulum (the Triangle), Argo Navis (the Great Ship), Ara (the Altar), Libra (the Scales or Balance) and Lyra (the Lyre). In Homeric astronomy Corona Borealis is equivalent to the shield of Odysseus, Corona Australis to Pandarus' quiver and Argo Navis to the shield of Great Aias, while Ara is the table upon which Hecamede sets a feast before Patroclus and King Nestor. Crater, Lyra and Libra become the homes of the warriors Odius, Eurypylus and Polydamas. The brightest stars in the relatively undistinguished constellations of Sagitta and Triangulum are represented by unremarkable warriors who play no active part in the *Iliad*, such as Gouness of Cyphus and Dodona, and Nireus of Syme.

Less Important Stars and Warriors

• *Stars represented by warriors about whom Homer gives little information. These stars are usually in groups adjacent to their killers, or on unobstructed lines of sight from them.*

Identification through the Catalogue of Ships and through the Rule of Wounding are the most precise of the methods Homer uses to associate warriors with stars in constellations, but there

Table 4 *Stars in the catalogues of Homer and Ptolemy*

Modern catalogue	Homeric catalogue: warrior and wound	Ptolemy's *Almagest*
CONSTELLATIONS IN HUMAN FORM		
ε Orionis	Polydorus, speared in the belt by Achilles (20.407)	The middle of the three stars in the belt of Orion
γ Cassiopeia	Elephenor, speared in the flank by Agenor (4.463)	The star just . . . over the thighs of Cassiopeia
γ Persei	Diomedes, wounded in the right shoulder by Pandarus' arrow (5.98)	The bright star in the right shoulder of Perseus
ε Boötis	Odysseus, wounded in the upper hip by Socus' spear (11.438)	Star on the right thigh, in the apron or girdle of Boötes
ζ Geminorum	Sarpedon, wounded in the left thigh by Tlepolemus' spear (5.661)	Star under the left knee of the rear twin of Gemini[6]
θ and τ Geminorum	Glaucus, wounded in the arm by Teucer's arrow (12.389)	θ = Star in the left forearm of the advance twin of Gemini τ = star in the same upper arm
δ Geminorum	Thrasymelus, speared in the belly by Patroclus (16.463)	Star in the left groin of the rear twin of Gemini
β Ophiuchi	Machaon, wounded in the right shoulder by Paris' arrow (11.507)	The more advanced of the two stars on the right shoulder of Ophiuchus
Capella, α Aurigae and β Aurigae	Nestor's powerful shoulders (23.627)	α = the star on the left shoulder of Auriga called Capella β = the star on the right shoulder
CONSTELLATION AS PART MAN AND PART BEAST		
θ Centauri	Pyraechmes, speared in the right shoulder by Patroclus (16.287)	The star on the right shoulder of Centaurus
CONSTELLATIONS AS ANIMALS		
α Canis Minoris	Patroclus, speared by Hector in the lower belly or bowels (16.821)	The bright star called Procyon just over the hindquarters of Canis Minor
α Cancri	Diores, struck by Peiros on his right ankle with a stone, before being speared in the navel (4.517)	Star on the southern claw of Cancer
α Ursae Majoris	Agelaus, speared by Diomedes in the back (8.257)	Stars in the quadrilateral (the bowl of the Dipper) on the back of Ursa Major

Table 4 (*continued*)

Modern catalogue	Homeric catalogue: warrior and wound	Ptolemy's *Almagest*
γ Lupi	Pylaemenes, speared by Menelaus in the collarbone (5.576)	The rearmost of the two stars just over the shoulder blade of Lupus
ε Lupi	Scamandrius, speared by Menelaus in the middle of the back (5.49)	The star in the middle of the body of Lupus
χ Lupi	Peisander, struck by Menelaus in the forehead above the nose with a spear (13.611)	The rearmost of the two stars in the snout of Lupus
ψ^1 and ψ^2 Lupi	Peisander's eyes then fall out of his head (13.616)	The more advanced of the two stars in the snout of Lupus
α Cetii	Erymas, speared by Idomeneus in the mouth (16.345)	Three stars in the snout of Cetus. The rearmost, on the end of the jaw of Cetus
μ Tauri	Idomeneus helps a warrior wounded in the knee (13.212)	Star in the right knee of Taurus
Antares, α Scorpii	Menelaus' personal star in the 'heart' of the constellation from which Pandarus' arrow was deflected by Athene (4.185)	Three bright stars in the body of Scorpius. The middle one of these which is reddish and called Antares

CONSTELLATIONS AS INANIMATE OBJECTS

β Lyrae	Eurypylus, wounded by Paris in the right thigh with an arrow (warrior seen in profile) (11.583)	The northernmost of the two advanced stars on the bridge of Lyra
δ and γ Crater	Odius – Agamemnon's spear enters Odius' back and exits through his chest (5.39)	γ = the southernmost of the two stars in the middle of the bowl δ = the northernmost

are many lesser warriors about whom he gives little information and yet who still need a home in the heavens. Some victims are not identified individually but are referred to simply as, for example, 'Lycians'.

The principle for identifying the stars with which such warriors are associated is that generally they are less prominent stars on a 'line of sight' or close to their attackers in the heavens. Groups of Trojans killed in large numbers by

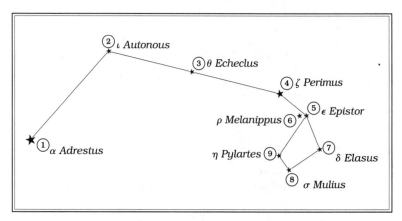

Fig. 18 The rhythm of the poetry divides nine stars in the neck and head of Hydra into three groups of warriors killed by Patroclus.

Patroclus and Achilles, for instance, are stars close to Canis Minor and Canis Major. The stars such lesser warriors represent belong to less prominent constellations or to clusters of stars in the outer boundaries of constellations. Haphazard as this method may at first seem, it is successful in determining the position of many stars. In Book 16 of the *Iliad* a group of nine Trojan warriors are killed by Patroclus in this sequence:

> First Adrestus, Autonous, Echeclus,
> Perimus, the son of Megas, Epistor and Melanippus;
> after these he killed Elasus, Mulius and Pylartes. (16.693)

These victims match stars – in a clockwise direction – in three groups in the neck and head of the adjacent constellation of Hydra (fig. 18). Patroclus kills the Trojans of Hydra in this order:

— 1 Adrestus, α (magnitude 2.0); 2 Autonous, ι (3.9); 3 Echeclus, θ (3.9) – three individual killings identify three widely spaced stars.

— 4, 5 and 6: Perimus, son of Megas, ζ (magnitude 3.1); Epistor, ε (3.4); and Melanippus ρ (4.4). All die in a flurry. Perimus, ζ (3.1), is distinguished from his two companions by being a 'son of . . .', and is allocated the brightest star of the three. Although the difference in magnitude between the stars of Perimus and Epistor does not appear to be numerically great, it is significant enough to be seen with the naked eye.[7]

— 7, 8 and 9: Elasus, δ (magnitude 4.2); Mulius, σ (4.5); and Pylartes, η (4.3) die in succession and their stars are of similar magnitude.

THE RULE OF SENIORITY

When Homer introduces groups of men of lower rank, he commonly singles out one of them for special emphasis, and that man represents the brightest star in his group. By applying this rule to the band of Trojans from Hydra, killed by Patroclus, we can see that Perimus, *son of Megas*, is indicated as the brightest of a group of three stars and is almost half a magnitude brighter than his immediate comrades.

In a second flurry of nine killings in Book 16, Patroclus 'laid low, one after the other, Erymas, Amphoterus, Epaltes; Tlepolemus, son of Damastor, Echius and Pyris; Ipheus, Euippus and Polymelus, son of Argeas' (16.414). The Rule of Seniority would suggest that these men would match a group of nine stars whose magnitudes were in the following sequence:

1. Erymas, Amphoterus, Epaltes – three stars of similar magnitude.
2. Tlepolemus, son of Damastor, Echius and Pyris – the first star significantly brighter than the other two.
3. Ipheus, Euippus and Polymelus, son of Argeas – the last star in the subgroup brighter than the other two.

101

For these men to have been killed by Patroclus suggests that their stars would have to be close to, or on an unobstructed line of sight from, α Canis Minoris, the personal star of Patroclus. Study of a star chart shows that such a group of stars with magnitudes in the correct sequence lies in Gemini, the constellation adjacent to Canis Minor. The nine stars in Gemini are:

— *First group*: Erymas, 85 (magnitude 4.9); Amphoterus, 81 (5.1); and Epaltes, 74 (5.0) – all of very similar magnitude.
— *Second group*: Tlepolemus, son of Damastor, λ (magnitude 3.6); Echius, 45 (5.2); and Pyris, 41 (5.0). Tlepolemus is considerably brighter than his companions.
— *Third group*: Ipheus, 38 (4.1); Euippus, ξ (3.4); and Polymelus, son of Argeas, γ (1.9). Polymelus is much brighter than his companions.

'Son of . . .' is not the only phrase attached to warriors to denote a brighter star. In Book 11, when Odysseus kills or wounds warriors representing stars in Serpens Caput, Homer describes Deiopites as 'noble' (11.420), and his star is the brightest, and in the correct place, in the sequence of attack.

Chariots, Spears and Corslets

• *Stars identified by artefacts and other devices. Spears, armour, chariots and bows are among the paraphernalia of war which identify many more stars.*

Boötes, the constellation of Odysseus (fig. 19), provides a typical example of how Homer used artefacts to describe stars. The shield, armour and cloak of Odysseus draw attention to Corona Borealis (the shield) and two groups of stars at the feet of Boötes, on one side where the armour was 'laid upon the ground' (3.195) and on the other where the squire Eurybates picked up his master's cloak (2.183). Another prominent

feature of Boötes is a line of stars that points towards celestial
north and which represents the staff with which Odysseus beat
Thersites (Coma Berenices), the miserable warrior who cast
doubt on the leadership of Agamemnon (2.265).

Pandarus' quiver, out of which he took an arrow to shoot at
Menelaus (4.116), is Corona Australis, a small group of stars
close to Sagittarius (fig. 20). Rectangles of stars within constellations are commonly referred to by Homer as the 'corslet' or
'shirt' of a warrior. An example can be seen in the drawing of
Cepheus (fig. 21), where four stars represent the linen breastplate (2.529) of Little Aias, set against the background of the

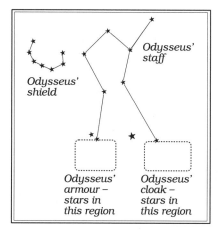

Fig. 19 Odysseus as Boötes.

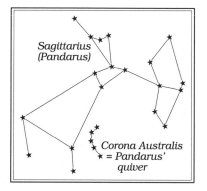

*Fig. 20 Pandarus' quiver identifies
stars in Corona Australis.*

Milky Way. Another item of clothing found in the sky is Achilles' helmet (18.610), a group of stars in Monoceros, the modern constellation above the 'head' of Canis Major (see fig. 46).

Throughout the *Iliad*, the chariots of warriors are represented by triangles of three stars. A good example is seen in fig. 22, which shows the two chariots of King Nestor that lead his troops into battle – 'He placed his knights with their chariots and horses in the front rank . . .' (4.297). Another group of stars in the triangular configuration of a chariot is found close to the

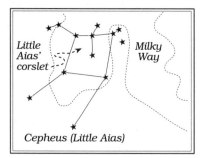

Fig. 21 *The rectangular breastplate of Little Aias in Cepheus. Cepheus is a circumpolar constellation with its 'foot' near to celestial north.*

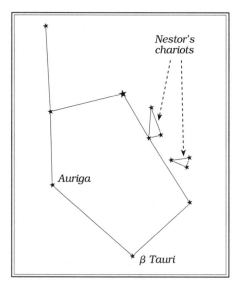

Fig. 22 *Two groups of three stars that make Nestor's chariots.*

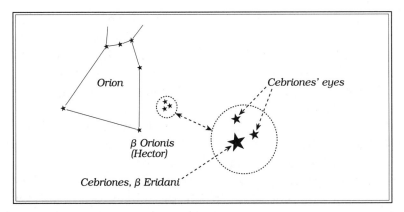

Fig. 23 *The enlarged section in the circle shows Cebriones and the two eyes which pop out of his head when he is killed.*

personal star of Hector in Orion. Homer draws further attention to this group in another manner when he tells how Cebriones, the charioteer to Hector, met his death when his brow was smashed by a stone and his eyes fell out on to the ground (16.739). Physically it would not be possible for someone's eyes to pop out and roll in the dust, but it is an unforgettable way of drawing attention to the personal star of Cebriones and two other small stars nearby.

To identify Cebriones' personal star it is necessary to use the Homeric logic that identifies stars that represent the aides, squires and heralds of commanders. These stars are always close to the personal stars of their masters, so, as Cebriones is an aide to Hector, his star will be a relatively bright star close to β Orionis, the personal star of Hector. As Cebriones was a charioteer, his star is also likely to be part of a triangular group of stars. Such a group lies just outside the constellation outline of Orion, and its brightest star – known today as β Eridani – is allocated to Cebriones. A closer look at that part of the sky (fig. 23) shows that β Eridani has two small adjacent stars – the eyes that pop from Cebriones' head. This is an example of how intertwined are the techniques of Homeric astronomy: not

105

only can that trio of stars be identified as a triangle-chariot, but Homer gives another reminder of its identity via the curious death of Cebriones. From its title of β Eridani, it can be seen that Cebriones' star is now part of the constellation of Eridanus, a long straggling constellation with few bright stars that wanders below the horizon deep into the southern hemisphere. The logic of Homer including it within the boundary of Orion can be seen in figs. 27 and 28.

The astronomical version of Cebriones' death may at first test the imagination, but Homer uses a similar method to identify three adjacent stars when describing the death of Peisander, χ Lupi, whose eyes, ψ^1 and ψ^2 Lupi, also fall out of his head when he is killed (13.617).

Women of Troy

Although their roles are not as decisive as those of the great warriors, the wives and companions of Agamemnon, Achilles, Hector and Paris are essential to the literary narrative. Like other mortals in the epic, these women have astronomical purposes as stars. In this section it will be shown how Homer gives two of them, Helen of Troy and Andromache, wife of Hector, places in the heavens.

HELEN OF TROY

A trawl of the *Iliad* for references to Helen reveals the following: Helen is the twin sister of Clytemnestra, wife of King Agamemnon; she is described as 'white-armed'; has a 'darling daughter'; is the daughter of Zeus; wears a white mantle; has two handmaidens in close attendance. Eventually, when the *Iliad* ends and Troy falls, Helen returns to live happily with her first husband, King Menelaus of Sparta.

To locate Helen in the sky from this information, a search began for two stars of similar magnitude to represent the twin sisters Helen and Clytemnestra, with their positions in the skies relatively close to those of their husbands, Menelaus (Scorpius) and his brother, Agamemnon (Leo). Libra provides just such a constellation with two bright stars of almost identical magnitudes, or 'twins': α^2 Librae has a magnitude of 2.7 and β Librae has a magnitude of 2.6. Libra is also within the zodiac near to both Scorpius and Leo. Although in modern times they are two separate constellations, Libra and Scorpius were once united in one large constellation, with the stars of Libra being known as the Claws of Scorpius. Exactly when the Claws were split off to form a separate constellation is not known, but the division of the larger constellation has echoes on Earth in the split between Menelaus and Helen when she was abducted by Paris. The two brightest stars of Libra thus make a good home for Helen (α^2 Librae) and Clytemnestra (β Librae), and Helen's daughter and servants are less bright stars close to α^2 Librae.

Other narrative from the *Iliad* which supports the case for Helen representing α^2 Librae includes:

1. Helen, 'with all her wealth' (3.91), is the daughter of Zeus (3.418). Her personal star, α^2 Librae, lies on the ecliptic – the apparent path of the Sun – a fitting place for a daughter of Zeus. A star on the ecliptic always has a position of status in Homeric astronomy.
2. Helen's 'darling daughter' (3.175), who is never named, suggests a star of a lesser magnitude. Very close to Helen's personal star, α^2 Librae, and visible to the naked eye, is α^1 Librae, with a magnitude of 5.2. Not only does this represent Helen's daughter, but it also gives Helen the appearance of being 'white-armed'.
3. When Helen dons a 'white mantle' (3.141) it is a reference to the Moon or a planet covering or occulting her personal star.
4. Helen's handmaids, Aethra and Clymene (3.144), are the stars μ Librae (magnitude 5.0) and ν Librae (magnitude 5.2), which are in 'close attendance' to α^2 Librae.

5. Helen survives the Siege of Troy and, in the *Odyssey*, returns to live with Menelaus; Scorpius and Libra once more live harmoniously as neighbours in the heavens.

ANDROMACHE AND ASTYANAX

Homer shows a compassionate understanding of the bonds of family love when Hector says farewell to his wife, Andromache, and his baby son, Astyanax, whom he knows he will not see again: '[Andromache] now came to meet him with a nurse who carried his little child in her bosom . . . Hector's darling son, and *lovely as a star*. Hector had named him Scamandrius, but the people called him Astyanax, for his father stood alone as chief guardian of Ilium. Hector smiled as he looked upon the boy, but he did not speak, and Andromache stood by him weeping and taking his hand in her own' (6.399). This scene, foretelling the destruction of Troy and its people, stands alone as a moment of tenderness during the fierce heat and slaughter of war.

An astronomical place can be found for Andromache in the skies. Homer says that Andromache's father and six brothers were killed when Achilles sacked Eëition; Andromache also has a nurse who carries her young son. From this information it was reasonable to look for a constellation close to Canis Major (Achilles) and Orion (Hector) that had eight brighter stars to represent Andromache, her father and six brothers. Such a grouping is found in the constellation of Lepus, at the foot of Orion, which has eight stars brighter than the fourth magnitude. Andromache is allocated to β Leporis, and her father to α Leporis, while the other brighter stars represent her six brothers. Two fainter stars are Astyanax and his nurse (fig. 24).

Techniques similar to those used for Helen and Andromache are used to place the companions of Achilles and Agamemnon, Briseis and Chryseis.

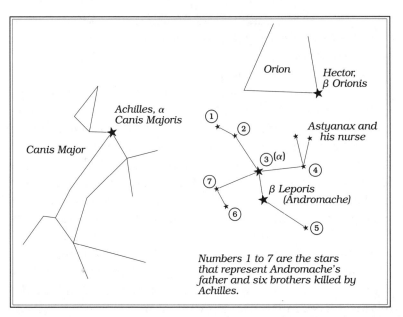

Fig. 24 *Andromache (β), her son and his nurse, together with her father (α) and six brothers, represent the brighter stars in the outline of the constellation of Lepus. The killing of Andromache's father and brother by Achilles, and her marriage to Hector, puts Lepus into a relationship with Canis Major and Orion.*

Creating a Chart of the Skies

Homer's identification of many hundreds of stars leads to the problem of how to place them in their correct locations in the heavens and so create a chart of the skies that can be retained in memory. The answer is again found in the killing fields of Troy, and is applied with fine Homeric logic.

Always, the killing of one warrior by another constitutes a line of sight as Homer directs us to look from a star in one constellation to a star in another constellation. The most powerful warriors – those who do most of the killing – have their personal stars in such well-known constellations as Leo (Agamemnon), Scorpius (Menelaus), Perseus (Diomedes), Gemini (Sarpedon), Taurus (Idomeneus) and Orion (Paris).

When their spears and arrows fly across the sky and strike warriors from other regiments, Homer thus identifies stars in other constellations. Many victims are men of little consequence and represent less prominent stars in their constellations. Their purpose is valuable, however, in that they allow Homer to expand his chart of the heavens beyond the constellations of the major players. The ancients who were intimate with the detail of such killings would have been equally familiar with which configuration of stars represented each of the forty-five regiments of Greeks and Trojans. As weapons flew from a warrior-star in one constellation towards a warrior-star in another, the relative position in the skies of each star and its constellation, from the best known to the least known, was reaffirmed in Homer's conceptual chart of the heavens.

An example from Book 11 shows how Homer uses an incident during a furious assault on the Trojan lines to establish the relative places in the heavens of stars in the constellations of Orion, Lyra and Ophiuchus. To try to stem the Greek assault, Paris of Troy wounds the Greek leaders Machaon (11.506) and Eurypylus (11.583) with arrows from his bow (fig. 25).

The first warrior to be hit by Paris (α Orionis) is the healer Machaon, a son of the physician Aesculapius. Since ancient times, the constellation Ophiuchus, around which are wrapped the constellations of Serpens Caput (the Serpent's Head) and Serpens Cauda (the Serpent's Tail), has been identified with Aesculapius. Even today, his snake-entwined staff is still used as a medical symbol. Machaon, wounded in the right shoulder, is placed at β Ophiuchi, a star in the 'right shoulder' of Ophiuchus. When Paris shoots his arrow and wounds Machaon, their personal stars, α Orionis and β Ophiuchi, and the constellations of Orion and Ophiuchus are fixed in memory on Homer's sky chart.

The second warrior wounded by Paris is the Greek leader Eurypylus (Vega, α Lyrae), in the right thigh (β Lyrae). Paris' arrow establishes a link between the constellations of Orion

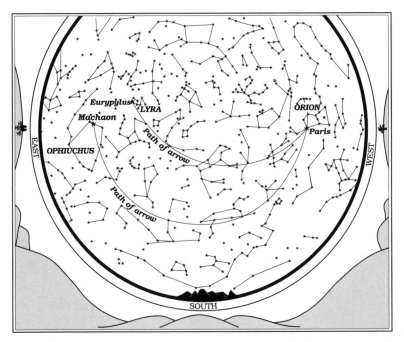

Fig. 25 *The Trojan prince Paris (α Orionis) fires arrows at Machaon (β Ophiuchi) and Eurypylus (α Lyrae).*

and Lyra and between the stars Betelgeuse (α Orionis) and Vega, the brightest star in the northern hemisphere. An associated dramatic effect linked to Lyra is the Lyrids meteor shower; this is represented by the blood that drips from Eurypylus' wound. In these two instances, Homer also gives two more views of the celestial sphere between different points of the horizon: Orion to Ophiuchus and Orion to Lyra. The furious fighting that continues in Book 11 gives possibly the most comprehensive conceptual chart of constellations of the night sky in the entire *Iliad*. Constellations identified with Greek warriors are Leo, Perseus, Argo Navis, Lyra, Eridanus, Ophiuchus and Boötes, and those of the Trojans are Ursa Major, Virgo, Orion, Aquarius, Equuleus and Serpens Caput.

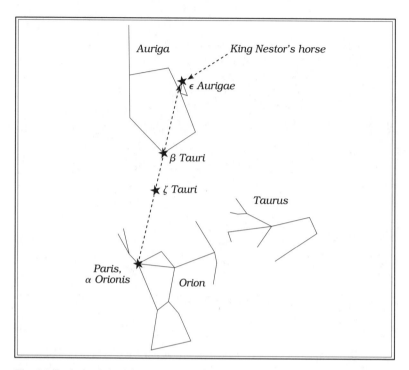

Fig. 26 *Paris (α Orionis) wounds King Nestor's horse (ε Aurigae) in the Kids, by lining it up with the stars ζ Tauri and β Tauri.*

Warriors are not the only victims of the deadly weapons that hurtle across the sky. The scholar E. V. Rieu wrote that more readers would be brought closer to tears by the death of a horse in the *Iliad* than by the killing of all of Penelope's suitors in the *Odyssey*.[8] The death of a horse harnessed to the chariot of King Nestor is not only an emotive incident but another example of how narrative makes memorable connections between constellations and stars. '. . . one of his [Nestor's] horses was disabled. Paris . . . had hit it with an arrow just on the top of its head where the mane begins to grow away from the skull, a very deadly place. The horse bounded in his anguish as the arrow pierced his brain, and his struggles threw the others into confusion' (8.80).

With the flight of one arrow from the bow of Paris, four bright stars are connected by a straight line: α Orionis, ζ Tauri, β Tauri, and ε Aurigae (fig. 26). The arrow strikes Nestor's horse, ε Aurigae, 'on top of the crown', the star at the apex of the triangle of stars known as the Kids. This is a variable star whose fluctuation is visible to the naked eye, and its changes in magnitude are represented by the struggles of the dying horse.

Nestor's is not the only horse that identifies a star. In Book 5, Pandarus bemoans that he took no horses with him to Troy, even though 'In my father's stables there are eleven excellent chariots, fresh from the builder, quite new, with cloths spread over them; and by each of them there stand a pair of horses, champing barley and rye' (5.193). This short passage is a testimony to the observational skills of ancient astronomers and the precision with which they could determine the brightness or magnitude of stars. The eleven pairs of horses to which Homer refers are twenty-two stars in Sagittarius (the astronomical home of Pandarus and his father) and contiguous Corona Australis (Pandarus' quiver). The 'barley' and 'rye' which they are eating are common Homeric codewords and indicate that the stars are set against the background of the Milky Way; the chariots cannot be seen, for they are covered with cloths. It will never be known who made the first study of this corner of the sky, but they were expert enough to observe that twenty-two stars could be 'paired' according to their brightness or magnitude. Table 5 (*overleaf*) lists twenty-two stars that match Homer's description.

Achilles has a trio of fine horses, and one of them, Xanthus, is the magical animal that warns him that his days are numbered (19.404). In fact two of the horses, Xanthus (δ Canis Majoris) and Balius (η Canis Majoris), are immortal and mark stars in the Milky Way. Pedasus (ε Canis Majoris) is 'mortal' and, although said to be every bit as good as the immortal pair (16.149), his star is just outside the Milky Way. The configuration of these makes the familiar triangle of one of Homer's chariots.

113

Table 5 *Paired stars in Sagittarius and Corona Australis*

Sagittarius
1.   ε (1.9) and σ (2.0)
2.   ζ (2.7) and δ (2.8)
3.   λ (2.9) and π (3.0)
4.   γ (3.1) and η (3.2)
5.   φ(3.3) and τ (3.4)
6.   ξ (3.6) and o (3.9)
7.   μ(4.0) and ρ¹ (4.0)
8.   α (4.1) and β (4.1)

Corona Australis
9.   γ (4.1) and ε (4.1)
10. α (4.0) and β (4.0)
11. δ (4.6) and ζ (4.7)

A long-standing puzzle in the *Iliad* has been why Homer sometimes refers to horses as being 'single-hooved' or 'single-footed' (5.829, 8.157, 9.126 etc.). An ingenious literary explanation has proposed that horses are 'single-footed' in that they do not have cloven hoofs.[9] A more likely explanation is to be found in the skies, with ancient Greek observers comparing the movement of stars across the heavens with horses galloping across the field of battle – 'single-hooved' horses indicating single stars. Nowhere does Homer use the epithet to greater effect than when Achilles, dressed and armed for battle, leads out his 'single-hooved' horses as the stars of Canis Major rise above the horizon (19.424). Nor is the comparison of horses with stars confined to the *Iliad*: in parts of more northern Europe, stars were believed to be steeds tethered to the pole star.[2]

The Boundaries of Constellations

Allocating warriors to represent brighter stars in the outlines of constellations is relatively straightforward when following Homer's rules on Wounding, Magnitude and Seniority. Others may challenge our interpretation of where a warrior's personal star is in the sky, but we have never identified a star in *Homer's Secret Iliad* simply in order to eliminate a warrior from Homer's list.

A problem for ancient astronomers was to identify the numerous stars that lie between constellation outlines. These generally fainter stars were not named individually, and Ptolemy, in the *Almagest*, solved the difficulty by describing such groups as 'Stars around [such-and-such constellation]'. Homer overcame the problem in a not too dissimilar way. For example, lying between Leo (Agamemnon) and Ursa Major (Troy) lie many faint stars in the modern constellations of Leo Minor and Lynx. So undistinguished are the stars of Lynx that Hevelius, in the seventeenth century, gave the constellation that name because he felt the observer would need the 'eyesight of the lynx' in order to see it at all. Homer portrays these stars as unnamed and uncounted warriors killed by Agamemnon as he fiercely 'went about attacking the ranks of the enemy with spear and sword and with great handfuls of stone' (11.265). The poorly defined group of stars that lies between Cygnus and Cassiopeia was in AD 1690 defined as the constellation of Lacerta, the Lizard. To Homer, these stars become the anonymous Trojan warriors killed in a flurry of fighting by Little Aias (14.520).

A glance at fig. 27 will show how complex constellation boundaries have become in modern times, but nevertheless the *Iliad* gives an occasional clue about the boundaries that Homer envisaged for at least a small number of constellations. Indications of these are seen in his references to the 'huts' and 'tents' of leading warriors who have been assigned via the Catalogue of Ships to stars within those constellations.

Homer's narrative indicates that stars close to modern boundaries were not always in the same constellations that they are today. A good example is El Nath, β Tauri, now placed at the tip of the northernmost horn of Taurus, the Bull. Homer, like Ptolemy, shared this star between Taurus and Auriga.

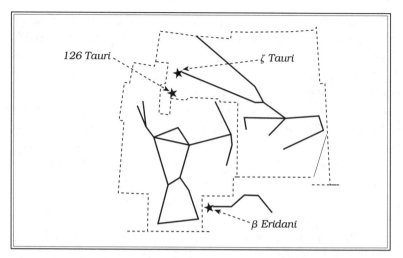

Fig. 27 *The dotted lines show the complicated modern boundaries of Orion, Taurus and part of Eridanus.*

Similarly, α Andromedae is shared between the constellations of Andromeda and Pegasus, where it forms one of the corners of the well-known 'square' (see fig. 14). Homer's constellation boundaries would of necessity have been much simpler than modern ones, and a projection of how he might have seen the constellation of Orion is shown in fig. 28. Two warriors who, as sons of Priam, have their Homeric home in Orion are today placed in Taurus. These are Deiphobus, ζ Tauri, and Antiphus, 126 Tauri, who are placed according to the Homeric Rule of Wounding, while Hector's charioteer, Cebriones (today placed at β Eridani), is as close to his master in Orion as an aide should be. It will be seen in the profile of Thetis and her nymphs in Chapter 6 that the logical allocation of β Eridani to Homer's Orion is no loss to Eridanus.

Homer's 'huts' and 'tents' are areas of sky within each of which lie constellations represented by warriors. In the case of Taurus, Homer indicates the extent of the eastern boundary of the constellation by saying that Idomeneus keeps a 'bundle

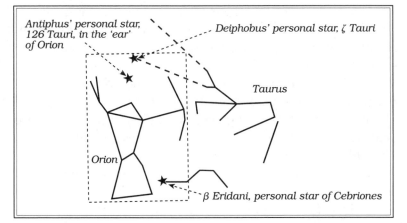

Fig. 28 *This diagram shows how the Greeks of Homer's era may have seen the boundaries of Orion.*

of spears' resting against the wall of his hut (13.260) – a metaphor for the group of stars known as the Pleiades. It will be shown in Chapter 7 that the 'hut' of Achilles (α Canis Majoris) included the stars of Patroclus (α Canis Minoris) and Phoenix (β Cancri). Chapter 7 will also reveal how reclassification of the heavens occurred in ancient times as well as more recently.

The Rule of Magnitude

As we saw in Chapter 1, in modern times stars are identified not only according to their place within a constellation but also according to their relative brightness or magnitude; and so it was in Homer's day too. On the modern scale of stellar magnitude, the stars in one division are about 2.5 times brighter or fainter than those in the next highest or lowest division, so that a star of magnitude 1 is 100 times brighter than one of magnitude 6. Each division is further divided into ten subdivisions: for example, Rigel, β Orionis, has a magnitude of 0.1, while Betelgeuse, α Orionis, is a little less bright with a magnitude of 0.5.

We have already seen how Homer used rank as a way of comparing the brightness of stars *within* constellations – a commander will be brighter than a warrior of the second rank, and so on down the line of command to insignificant warriors or less prominent stars. There still remains the difficulty of comparing the brightness of stars in different constellations, and Homer achieves this through the Rule of Magnitude: a warrior (star) can be killed only by a stronger warrior (brighter star). This rule means that, as Sirius is the brightest star in the sky, Achilles cannot be killed by any other warrior. And indeed Achilles is still alive when the *Iliad* ends (though he is dead when the *Odyssey* begins). Nowhere does Homer describe an astronomically impossible event, and nor does he say who killed the great hero.

By applying the Rule of Magnitude, Homer compares the brightness of many stars across the entire heavens. For example:

— If Warrior A kills Warrior B, then Warrior A is a brighter star than Warrior B.
— If Warrior C kills Warrior A, then Warrior C is brighter than Warriors A and B.
— If Warrior D kills Warrior C, then Warrior D is brighter than Warriors A, B and C.

In the *Iliad* these conclusions can be seen in the deaths of three leading warriors and the actions of Achilles:

1. Patroclus (A) kills Sarpedon (B).
2. Hector (C) kills Patroclus (A).
3. Achilles (D) kills Hector (C).

Homer has established an order of magnitude based on Patroclus (Procyon, α Canis Minoris) being brighter than Sarpedon (Pollux, β Geminorum), Hector (Rigel, β Orionis) being brighter than Patroclus, and Achilles (Sirius, α Canis Majoris) being the brightest of all. Modern observations confirm what was deduced by ancient observers: the magnitudes concerned are Achilles -1.4, Hector 0.12v, Patroclus 0.4 and Sarpedon 1.1. To the naked eye, the difference in magnitude between Hector and his victim, Patroclus, is not great, and its recording in such a positive manner shows the observational skills of Homer and his predecessors.

It is not known what system Homer devised to compare the magnitudes of stars, but it is possible to determine the brightest stars by observing the order in which they become visible in the sky at dusk. The brightest stars are the first to appear; then, as the dark of night increases, stars of fainter magnitude – lowly warriors – appear with a rush, ready to be killed by the stronger warriors already in the sky and waiting to strike (fig. 29).

Red-Haired Warriors and Cowards

After some important warriors have been placed in their constellations, Homer reaffirms the visible qualities of their stars by the ingenious use of epithets and similes. These are usually considered to be literary devices dictated by considerations of poetic metre, and have sometimes been omitted by translators. In the preface to his translation of the *Iliad*, Samuel Butler says, 'had Homer written in prose he would not have told us these things so often'. Epithets, however, carry important astronomical information about stars.

The most obvious attribute of any star is its relative brightness, and for Achilles, the most powerful warrior at Troy, Homer uses such epithets as the 'mightiest of all the Achaeans' (19.216) and 'above his peers' (1.505). These apply not only to the great hero, but also to his personal star, Sirius, the brightest star in the heavens. Achilles is also a 'fast runner' or is said to have 'swift feet' – phrases which draw attention

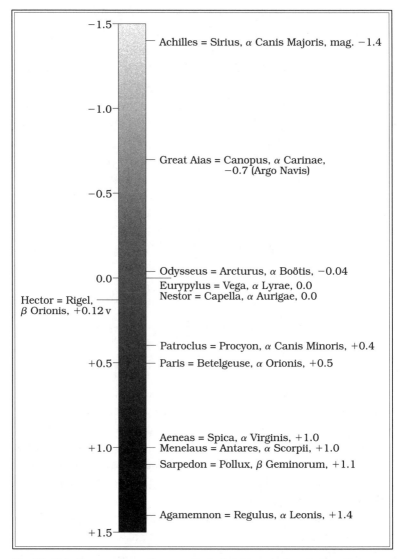

Fig. 29 *This diagram shows the magnitude (brightness) of a selection of stars and the warriors with whom they are associated. They are in descending order of brightness, from Sirius (Achilles), the brightest star, to Regulus (Agamemnon), which is the twenty-first brightest.*

to the relatively short time it takes Sirius to cross the sky from rising to setting compared with constellations higher in the sky.

'Wide-ruler' (1.102) and 'King of Men' (1.130), Agamemnon of golden Mycenae, the commander-in-chief of the Greeks, is Regulus (α Leonis, magnitude 1.4). Although only the twenty-first brightest star, Regulus gains its great status by being the brightest on the ecliptic, the apparent path of the Sun. The name 'Regulus' is a diminutive of Rex (King), and in Babylonia, too, the star was named Sharru, the King.[11] Homer exults in describing Agamemnon in the glittering terms which befit his standing, particularly when Agamemnon as α Leonis rises at one time of year at the same moment as the Sun and travels across the daytime sky covered or occulted by the brightest object in the skies (11.15–283).

The leading Trojan warriors, Hector (Rigel, β Orionis) and his brother Paris (Betelgeuse, α Orionis), represent two stars that can vary in brightness, and have roles attached to their characters that reflect this. Hector has strange periods when he temporarily retires from battle, only to return to the fighting rejuvenated. Hector's star, β Orionis, is variable over the narrow range of 0.08 to 0.2, but usually shines at magnitude 0.01. In another incident, Hector was 'as some baleful star shines for a moment through a rent in the clouds and is again hidden beneath them' (11.62). When Hector is attacked by Aias in their duel in Book 7 he is knocked down and brought to his feet only with the help of the god Apollo (7.271). In Book 3, flamboyant but cowardly Paris twice escapes to safety from duels, as is explained in Chapter 7. His fluctuation in temper-ament is matched by his personal star, α Orionis, which varies in brightness from magnitude 0.1 to 0.9 – a wider range than that of his brother. Glorious Hector of the 'flashing helmet' (6.263, A. T. Murray translation) is also the warrior who glitters 'like a snowy mountain' (13.754), and this may bring to mind the image of a shower of meteors around the head of Orion, where, in present times at least, the annual Orionids meteor shower originates.

Stars vary not only in brightness but also in colour, and sharp eyes can easily see that a number of them are distinctly red. Idomeneus represents the red star Aldebaran, α Tauri, and Homer describes him as 'looking like a god' (3.230) and like a 'flame of fire' (13.330). Idomeneus' hair is described as 'already flecked with grey' (13.361), probably a reference to faint stars to the north of Taurus. And Idomeneus' eyes are not of 'the sharpest' (23.476) – a reflection, perhaps, of the dull-red characteristics of Aldebaran. Ares is the planet Mars, which, like Aldebaran, is red and in the zodiac. The personal star of Menelaus is another bright red star, Antares, α Scorpii, and the warrior from Sparta is referred to as 'red-haired' and a peer of Mars, the red planet.

5

Warriors as Constellations

> With a unique combination of brashness
> and reverence, the Greeks transformed the
> universe into a picture book.
>
> Rudolf Thiel, *And There was Light* (1958)

Since the days of antiquity, observers of the skies around the world have arranged stars into patterns, and always for the same reasons – to bring order and understanding to the heavens. Some groups of stars are so striking that they have been seen in a similar way in cultures widely separated in distance and time. A common strand of human thought is seen, for instance, in the way in which three different societies perceived the shape of the seven best-known stars of Ursa Major that are popularly known today as the Plough or Big Dipper (fig. 30). In Homeric astronomy they represent the wain or wagon on which King Priam took Hector's body back to Troy. Edna Leigh wrote, 'Among the Arabians, the Big Dipper represents a bier and mourners. The four stars which form the wagon or cart represent the bier, and those in the handle are the mourners'.[1] A similar funereal image was seen by the Sioux, native North Americans.[2]

In more imaginative eras than now and under the influence of great literature, constellations have been seen in the form of mighty heroes, animals, religious figures and mythical beasts. In modern scientific times, constellations are depicted as austere patterns of straight lines, but even so the various authorities do not always agree on how a constellation should look in all details.

Homer had an ingenious way of identifying constellations that was admirably suited to the technique of preserving astronomical learning in narrative. It was noted in Chapter 4 that, although all of Homer's warriors had 'personal' stars, some of the more significant leaders could also represent an entire constellation. This became apparent when we searched the *Iliad* for narrative that might match warriors to their personal stars. Such was the surfeit of information about Agamemnon, Paris, Hector, Odysseus, King Nestor and Diomedes that it was felt that Homer had a second purpose for these leaders. It was a remarkable moment when we compared the descriptive passages to the constellations in which leaders had their personal stars. It became clear that, although Homer used similar configurations of constellations to those commonly thought to have first been arrived at in Mesopotamia, he superimposed upon them the figures of warriors in the *Iliad*. Narrative that described the Deeds of Glory or *'aristeia'* of Agamemnon, Achilles and Diomedes was thus allegory that related to the familiar constellations of Leo, Canis Major and Perseus – three

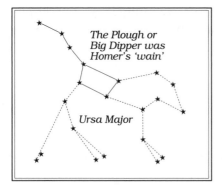

Fig. 30 The seven stars which make up the Plough or Big Dipper within Ursa Major are linked by solid lines in this diagram.

of the best-known groups in the heavens. Narrative about Paris, Hector, Odysseus and King Nestor also related these characters to constellations, supporting this dramatic extension of Homeric astronomy.

Sometimes the link with a constellation is quite straightforward; for example, Homer's many references to Achilles' heart within his 'shaggy breast' (e.g. 1.189) relate directly to the location of his personal star, Sirius, at the heart of Canis Major amid other stars. Agamemnon, on the other hand, has a 'black heart' (1.103), and, while that may say much about his character, a glance at a star chart shows that the 'heart' of Leo is in an area of dark sky. Odysseus represents the constellation of Boötes, and, as we shall see, when he is wounded in the hip, attention is drawn to the star Izar, ε Boötis. A cloak lying at the feet of Odysseus, picked up in Book 2 by his servant Eurybates, identifies stars at the foot of Boötes. Another group of stars lying close to the other foot of Boötes is the armour which Odysseus discards upon the ground in Book 3 (see fig. 41).

We shall see that King Nestor is also well served by Homer, who, in the process, twice describes in detail the constellation of Auriga. When Agamemnon, the Greek commander-in-chief, reviews his forces in Book 4, Nestor tells how he is going to deploy his regiment for battle, and draws in the mind a picture of all the stars of Auriga. Information about Nestor gleaned from Books 1, 8, 10, 14 and 23 shows his constellation in another light. Homer describes Nestor's physical attributes, from his shoulders, broad chest but weak legs and arms, to his clothing, footwear and shield. Among other leaders who merit attention as constellations is red-haired and broad-shouldered Menelaus of Sparta; when Pandarus wounds him in the stomach (4.134), blood trickles down his leg and collects as a pool of stars at his feet. This matches the broad-shouldered and spindle-legged constellation of Scorpius and its brightest star, the brilliant red Antares.

If further confirmation were needed that warriors could represent constellations, it came in an astronomical interpretation of the View from the Wall, a well-known incident in Book 3, when King Priam stands on the walls of Troy and asks Helen to point out to him the leaders of the Greek army (3.161). The apparent absurdity of a king who has been fighting a war for ten years not knowing the leaders of the opposing army has been a long-standing topic of discussion for scholars. The answer to this literary conundrum is found among the constellations represented by Greek commanders. From the questioning of Helen by Priam it is learned that:

— Agamemnon is taller by a head than Odysseus.
— Odysseus is broader across the shoulders than Agamemnon.
— When 'all are standing', Menelaus with his broad shoulders over-tops the entire company.
— Odysseus is 'more imposing' than Menelaus when they are seated.
— Great Aias is the bulwark of the Greeks.
— Idomeneus, leader of the Cretans, is like a god and surrounded by his captains.

Homer uses this scene to compare the sizes of four of the five constellations represented by Greek leaders (fig. 31). When the constellations they represent are compared on the same visual scale, his words take on a stunning new dimension. Idomeneus, the fifth leader named by Helen, is identified in a different manner. The leader of the Cretans has been identified as a peer of the god Ares (the planet Mars) (13.500), which not only places his personal star in the zodiac with the gods, but also indicates that it is a red star. Aldebaran, α Tauri, which represents Idomeneus, is the notably red star that marks the eye of the bull in Taurus. Helen says that Idomeneus is surrounded by his 'captains', just as Aldebaran is surrounded by

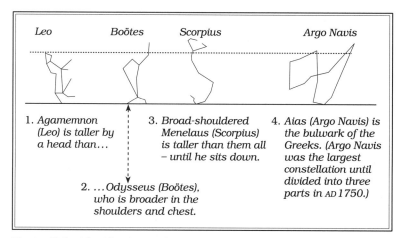

Leo	Boötes	Scorpius	Argo Navis

1. Agamemnon (Leo) is taller by a head than...

2. ...Odysseus (Boötes), who is broader in the shoulders and chest.

3. Broad-shouldered Menelaus (Scorpius) is taller than them all – until he sits down.

4. Aias (Argo Navis) is the bulwark of the Greeks. (Argo Navis was the largest constellation until divided into three parts in AD 1750.)

Fig. 31 Helen identifies the Greek commanders.

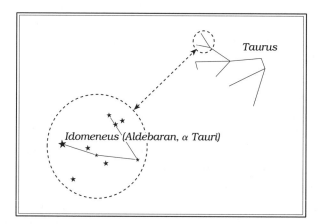

Taurus

Idomeneus (Aldebaran, α Tauri)

Fig. 32 Idomeneus is Aldebaran, α Tauri, and it can be seen in the enlarged part of the diagram that his star is close to those of his captains – the Hyades star cluster.

the Hyades, a well-known group of stars in that constellation (see fig. 32).

A full list of the Greek and Trojan commanders and their personal stars is given in Tables 7 and 8 at the end of this chapter.

Some of the constellations referred to by Homer as warriors are matched with the current constellation names in Table 6.

Table 6 *Homer's identification of constellations*

Warrior	Constellation
Achilles	Canis Major
Aeneas	Virgo
Diomedes	Perseus
Hector/Paris	Orion
Idomeneus	Taurus
Nestor	Auriga
Odysseus	Boötes
Patroclus	Canis Minor

The question that now has to be asked is, Why were some warriors identified with constellations as well as with the personal stars that are a part of the Homeric catalogue examined in Chapter 4? An answer can be found in the great set pieces of the *Iliad*, such as the duel scenes in Book 3, the wounding of Menelaus by Pandarus, Hector's journey to find his brother, the arming of Achilles, and the death of Hector. As we shall see, in these, Homer not only describes the movement of constellations across the sky but also explains his theories of the precession of the equinoxes and of the Earth's place at the centre of the universe. To have tried to follow individual stars as they crossed the heavens would have been difficult, but by giving warriors the shapes of constellations he could more easily describe their journeys – and his theories. It would have been awe-inspiring, in ages long past, when a poet-singer recited from the *Iliad* and his audience either watched or drew from memory images of the constellations playing out the drama as they wheeled across the heavens.

During this tranche of research, examination of the roles of Hector and Paris as constellations led to a curious result. Narrative from throughout the *Iliad* gave each man attributes that in certain events could identify him as representing the entire constellation of Orion, rather than just his personal star. The idea that two men could represent one constellation was at first troublesome, but closer examination of the narrative provided an explanation.

Orion is the home of the personal stars of King Priam's fifty sons, of whom Hector and Paris are the most famous. Orion, however, has a separate and equally important role in Homer's exposition of the nature of the universe, particularly when he seeks to describe how the celestial sphere rotates daily around the Earth. Homer's use of Orion in this way is found in Book 3, in the duel scenes between Paris and Menelaus (see Chapter 8, pages 224–5), in Book 6, when Hector is reunited with his brother (see Chapter 8, pages 229–31), and in Book 22, when Hector is killed by Achilles (see Chapter 3, pages 72–7). These incidents would have been even more dramatic to ancient audiences when seen as events in the sky, with Hector and Paris representing not just the pinpoints of light of their personal stars but the entire constellation of Orion.

Perplexing as this might at first seem, the narrative shows that Homer is very careful to ensure that there is no confusion about which of the brothers is representing Orion at any one time. When either Hector or Paris represents Orion in one of the great set pieces describing the movement of the heavens, the other is never involved in the same scene. When the two brothers do appear together – as when Paris tells Hector to return to the battlefield and he will follow on (see Chapter 8, page 230) – it is clear that they represent only their personal stars. In the pages that follow we show how a vivid extract of narrative describes Paris as a Trojan warrior, while in astronomical terms it gives an equally accurate portrayal of Orion.

There is a second reason for depicting a number of commanders as constellations. Few other named warriors are linked to the regiments of such leaders as Agamemnon, Menelaus, Odysseus and Diomedes, and another way had to be found to identify brighter stars in these regimental constellations. This was done by associating stars with the wounds suffered by these commanders. For example, when Pandarus wounds Menelaus in the region of his belt (4.135), the star thus identified is ε Scorpii in the 'belt' of Scorpius. Before being deflected by the goddess Athene, the arrow had been aimed at the 'heart' of Menelaus – the star Antares, α Scorpii, at the heart of the constellation. Similarly, when Diomedes is wounded in the flat of his 'right foot' the star identified is ζ Persei, in the 'foot' of the constellation.

Lest it be thought the idea of personifying constellations was unique to Homer, it should be noted that the Venerable Bede (AD 672–735) portrayed them as saints, and the seventeenth-century mapmaker Julius Schiller turned the zodiacal signs into the twelve apostles and the other constellations into Old and New Testament figures.

In the following pages, quotations are used to construct astronomical images of Orion, Auriga, Boötes, Perseus and Canis Major from narrative involving Paris, Nestor, Odysseus, Diomedes and Achilles. In the Prologue to *Henry V*, Shakespeare urged his audiences to use their imagination:

> Think, when we talk of horses, that you see them
> Printing their proud hoofs i' th' receiving earth;
> For 'tis your thoughts that now must deck our kings.

Homer makes similar demands when the *Iliad* is played on the stage of the universe.

Paris as Orion

The easiest of all astronomical interpretations of a warrior as a constellation is in Book 3, when Paris, the Trojan braggart, strides out from the ranks and challenges any of the Greeks to single combat in an attempt to end the war. Great is his terror when the gauntlet is picked up by Menelaus, cuckolded husband of Helen, whom Paris had spirited away to Troy. In terms of the earthly scenario of the *Iliad*, Paris' challenge creates a memorable picture of the Trojan in all his military grandeur (fig. 33). In terms of the celestial scenario, Homer is describing the splendours of the magnificent constellation of Orion (fig. 34). 'Paris sprang from the Trojan forward ranks, a challenger, lithe, magnificent as a god, the skin of a leopard slung across his shoulders, a reflex bow at his back and battle-sword at hip and brandishing two sharp spears tipped in bronze' (3.15, Fagles translation).

1. 'Magnificent as a god' = stars in Orion, in the crest of Paris' helmet, lie in the zodiac, home of the gods.
2. 'Skin of a leopard slung across his shoulders' = a group of stars below the area of the 'shoulder' of Orion are the spots of the leopard skin.
3. 'Reflex bow at back' = the 'bow' of Orion.
4. 'Battle-sword at hip' = those stars still known as the 'sword' of Orion.
5. 'Two sharp spears tipped in bronze' = two bright stars at the tip of spears held in Orion's right hand.

It is clear from Homer's description of Paris that he is facing left – or towards the east – confronting threatening Menelaus, who, as the constellation of Scorpius, is just appearing over the horizon.

Nestor as Auriga

There are strong mythical and astronomical links between Nestor and the constellation of Auriga. Known in Mesopotamia since ancient times as the Driver or Charioteer, conventional Greek legend associates this constellation with lame Erechtonius, King of Athens, who, crippled and unable to move about easily, invented the four-horse chariot and as a reward was placed by Zeus in the skies, as the constellation.

Fig. 33 *Prince Paris depicted as a Trojan warrior.*

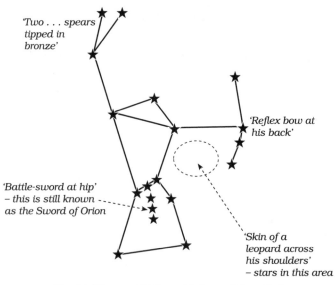

Fig. 34 *The constellation of Orion as Prince Paris.*

Homer uses elements of both sources to describe Auriga in the *Iliad*.

In the *Iliad*, King Nestor, the ageing ruler of Pylos, is known as the Gerenian charioteer and, like Erechtonius, he has a strong body but weak legs. Nestor's personal star is α Aurigae, the beautiful yellowish star known as Capella, or the She-Goat, whose offspring are two stars below and known as the Kids. This introduces another legend in which the infant Zeus was fed milk and honey by the horn of a she-goat; Homer tells us that words fell from the lips of the garrulous Nestor 'sweeter than honey' (1.247). Unfortunately, so the story continues, the tip of the goat's horn broke, but it was replaced by the nymphs so that Zeus could continue to feed; this mythical event gave us the phrase 'horn of plenty'. The clearly variable star ε Aurigae, one of the Kids, may well represent the tip of the horn, in the sense that after it 'breaks', or is fainter, it is restored to full brightness. In the context of the *Iliad*, Homer identifies ε Auriga when Paris fires an arrow and strikes King Nestor's horse on the top of its head (see pages 112–13).

When the many passages of narrative referring to Nestor are examined, they can be seen to have two separate themes. The first relates to the constellation described in the physical form of Nestor himself, with his prominent shoulders, head, elbows, sandals, clothes, shields and spears (see figs. 35 and 36). When the drowsy king is awakened by Agamemnon in Book 10, there is an image of a warrior raising himself upon his elbow – just as Auriga itself appears to do when it rises over the horizon. The second theme concerns Auriga as Nestor's regiment, with his five captains designated as the constellation's five brighter stars, his cowardly troops in the centre and brave foot soldiers in the rear, all of whom are led across the sky by two 'chariots' at the front (figs. 37 and 38).

It has long been wondered why Homer describes Nestor's homeland in the Peloponnese as 'sandy Pylos', for the remains of the so-called palace of Nestor lie a little inland on top of a

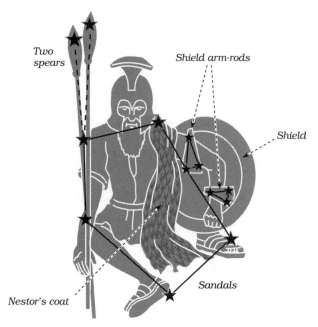

Fig. 35 *Auriga as King Nestor.*

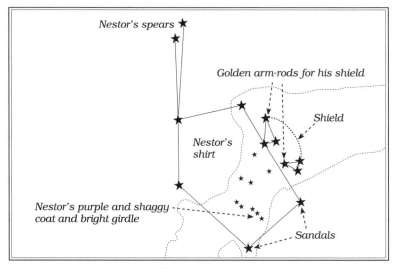

Fig. 36 *King Nestor and Auriga.*

rocky hill. The answer, as so often, lies in the heavens and the variety of ways in which Homer describes the Milky Way. In this view of Auriga, the Milky Way becomes the 'beach', the strand of light that rolls diagonally across the constellation, and whose myriad stars are as innumerable as grains of sand. Homer uses similar imagery later in the *Iliad* when the Milky Way adjacent to Canis Major is the 'seashore' upon which Achilles strolls when mourning the death of Patroclus. For a skymap of Nestor's homeland of Pylos, see Chapter 9.

> With this he put on his shirt, and bound his sandals about his comely feet. He buckled on his purple coat, of two thicknesses, large, and of a rough shaggy texture, grasped his redoubtable bronze-shod spear, and wended his way along the line of the Achaean ships. (10.131)

In this quotation Homer says that Nestor holds just one spear, but at 10.78 he describes the king as having two spears. Despite the apparent confusion, both quotations can be correct. In fig. 35, two stars are shown at the tips of the shafts of Nestor's spears, but usually only one of these, δ Aurigae, is included in the astronomical outline of Auriga (see fig. 37). Homer may well have been using the 'two spears' to draw attention to the fainter star, ξ Aurigae, which appears to be close by the side of δ. Nestor's shirt is the torso-like shape of Auriga; his sandals are two bright stars at the foot of the constellation; and the purple and shaggy coat is the dark and starlit areas of the constellation. Nestor also has a 'bright girdle', which is the Milky Way. Throughout the *Iliad*, when Homer uses certain 'codewords' they always have the same astronomical meaning. Shirts and corslets are groups of stars in a generally rectangular or pentagonal configuration, just as 'beach' and 'sandy' denote the Milky Way. The word 'purple' refers to a part of the sky with few stars, while 'shaggy' refers to areas with many stars. To suggest that Nestor had a purple and shaggy coat – i.e. that Auriga is bright and dark – is not the contradiction it might seem. The Milky Way is the 'shaggy' part of Auriga, and the 'purple' part is that which does not have many stars.

Further quotations from the *Iliad* referring to Nestor follow, together with the astronomical interpretations of the words:

— 'My son, all that you have said is true; there is no strength now in my legs and feet, nor can I hit out with my hands from either shoulder' (23.627). The abstract pentagonal shape of Auriga has no well-defined legs or arms, but it does have 'shoulders' and 'feet'. More references to Nestor's weak limbs occur in 4.313 and 11.667.

— 'Haste in pursuit, that we may take the shield of Nestor, the fame of which ascends to heaven, for it is of solid gold, arm-rods and all' (8.192). The arm-rods are the two groups of three stars that precede Auriga, and his shield is constructed from a semicircle of fainter stars around them (fig. 36).

— 'His [Nestor's] goodly armour lay beside him – his shield, his two spears and his helmet; beside him also lay the gleaming girdle with which the old man girded himself when he armed to lead his people into battle' (10.75). The various artefacts specified here can be seen in the drawings to match various aspects of Auriga.

— 'He [Nestor] raised himself [from his bed] on his elbow and looked up at Agamemnon' (10.80). In these few words, Homer firmly implants in the imagination the sight of Auriga rising above the horizon. As Auriga crosses the sky it adopts a more upright posture.

— 'Nestor was marshalling his men and urging them on, in company with Pelagon, Alastor, Chromius, Haemon, and Bias, shepherd of his people. He placed his knights with their chariots and horses in the front rank, while the foot-soldiers, brave men and many, whom he could trust, were in the rear. The cowards he drove into the middle, that they might fight whether they would or no' (4.293). In this scene, King Agamemnon is inspecting the troops and Nestor tells his commander how he will deploy his men in the forthcoming battles. In reality he is describing Auriga in the manifestation of his regiment rather than himself. Nestor's commanders are the bright stars in the outline of the constellation, his chariots are two groups of three stars, the cowards in the middle represent stars in the centre of the constellation, and his first-rate troops at the rear are stars that follow Auriga on its journey across the sky (figs. 37 and 38).

Odysseus as Boötes

Boötes as the Herdsman of the heavens is an ancient concept, and, in identifying this constellation with Odysseus, Homer acknowledges this pastoral association when he describes the Greek warrior stalking 'in front of the ranks as it were some great woolly ram ordering his ewes' (3.195). There are also geographical links between the shape of Boötes and the homelands of Odysseus in the islands of the Ionian Sea.

Odysseus excels in all manner of stratagems and cunning, and his reputation as a subtle and devious character and his later wanderings in the *Odyssey* are reflected in the 'wandering' of his personal star, Arcturus, α Boötis. This star has a noticeable 'proper motion', which means that over the centuries it has moved relative to other stars in the constellation. Since the time of Ptolemy it has moved from the 'thighs' of Boötes to the region of the lower leg. In astronomical terms, Arcturus has moved just over one degree, or a little more than the apparent diameter of the Moon.[3]

In Book 4 of the *Iliad* Agamemnon makes a tour of inspection of the regiments from Crete (Taurus), Locris (Cepheus), Salamis (Argo Navis), Pylos (Auriga) and Athens (Southern Cross), as well as those who sailed with Odysseus from

137

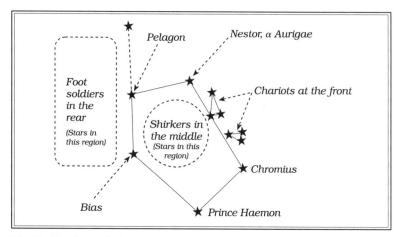

Fig. 37 *King Nestor's deployment of his captains, soldiers and chariots describes the constellation of Auriga.*

Cephallonia (Boötes) and with Diomedes from Argos (Perseus). For the commanders of most of these regiments, Agamemnon has encouraging words as he rallies his troops, but he scorns Odysseus, who, he says, would not join the fight even if he saw ranks of Greeks attacking the enemy in front of him. Given Odysseus' brave deeds in the rest of the *Iliad*, it is not obvious why Agamemnon should be accusing him of cowardice, until an examination is made of the part of the sky in which Boötes lies. There it can be seen that the constellation that precedes Boötes through the heavens is not one of the powerful Trojan constellations, but Coma Berenices (Thersites), an insignificant and faint Greek constellation. Such major Trojan constellations as Virgo, Gemini, Orion and Sagittarius are at a distance and can never confront Boötes. Agamemnon concedes this point, and before leaving he apologizes to Odysseus for his remarks. 'Odysseus [said Agamemnon] . . . excellent in all good counsel, I have neither fault to find nor orders to give you, for I know your heart is right, and that you and I are of a mind. Enough; I will make

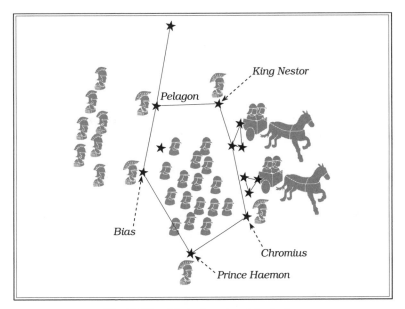

Fig. 38 *King Nestor's regiment in Auriga.*

you amends for what I have said, and if any ill has now been spoken may the gods bring it to nothing' (4.358).

In a second strange event that again casts doubt on the courage of Odysseus, he is accused by Diomedes of 'turning his back' on the fighting. 'Odysseus, he [Diomedes] cried, . . . where are you flying to, with your back turned like a coward?' (8.93). The astronomical connotation of this quotation relates to the passage of Boötes across the heavens. As Perseus, the constellation of Diomedes, rises, Boötes is already past the meridian and swinging around Homer's pole star (Thuban, α Draconis). Its attitude has so changed from the vertical that it appears to have turned its back on Perseus coming up behind (fig. 39).

The following quotations from the *Iliad* can be compared with the drawings of Odysseus the warrior and the astronomical diagram of Boötes in figs. 40 and 41.

139

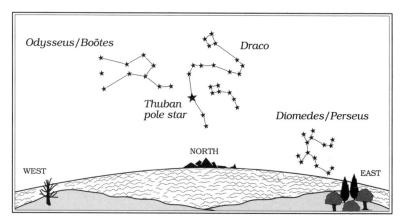

Fig. 39 *An observer looking north sees Boötes with its 'back turned' on the rising Perseus.*

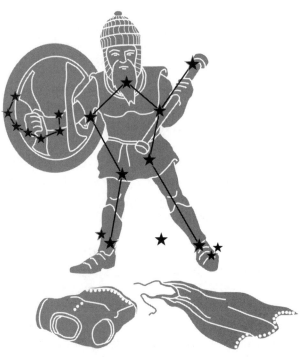

Fig. 40 *Odysseus drawn from narrative in the* Iliad.

— 'Whereon Odysseus went straight up to Agamemnon and received from him his ancestral, imperishable staff' (2.186). 'On this he beat him [Thersites] with his staff about the back and shoulders till he dropped and fell a-weeping. The golden sceptre raised a bloody weal on his back' (2.265).

— 'There was no play nor graceful movement of his [Odysseus'] sceptre; he kept it straight and stiff like a man unpractised in oratory' (3.218). One of the outstanding features of Boötes is the staff that points towards the north celestial pole and at the time in which the *Iliad* is set pointed towards the declining pole star Thuban, α Draconis.

— 'With these words he [Socus] struck the shield of Odysseus. The spear went through the shield and passed on through his richly wrought cuirass, tearing the flesh from his side'

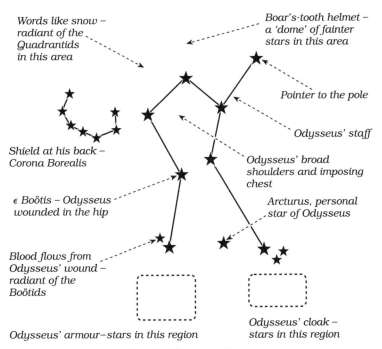

Words like snow – radiant of the Quadrantids in this area

Boar's-tooth helmet – a 'dome' of fainter stars in this area

Pointer to the pole

Odysseus' staff

Shield at his back – Corona Borealis

Odysseus' broad shoulders and imposing chest

ε Boötis – Odysseus wounded in the hip

Arcturus, personal star of Odysseus

Blood flows from Odysseus' wound – radiant of the Boötids

Odysseus' armour–stars in this region

Odysseus' cloak – stars in this region

Fig. 41 Odysseus as Boötes.

(11.434). Translations differ about where Odysseus was actually wounded, variously specifying the ribs, upper hip, entrails or bowels (11.437). The location of the wound may be Arcturus itself or even the star Izar, ϵ Boötis, close to which is a fainter star, 34 Boötis (magnitude 4.8), which could account for the flesh that is torn from the side of Odysseus. Izar itself is a double star, and with a low-powered telescope it can nowadays be seen to be of a yellow and green hue – the colour of a nasty and untreated battlefield wound. As an indicator of Odysseus' injury, Izar raises the question of whether the ancients had better than suspected techniques for naked-eye observations, or even if they had a form of optical aid (see Appendix 2).

A number of images are used to describe meteor showers in the *Iliad*. They include the tears of Agamemnon in Leo, Glaucus' bleeding arm in Gemini, Hector's flashing helmet in Orion, the flowing words of Nestor in Auriga, and the following quotation about Odysseus: 'When he raised his voice, and the words came driving from his deep chest like winter snow before the wind . . . no man thought further of what he looked like' (3.221). This simile indicates a meteor shower in the region of the Quadrantids close to the 'head' of Boötes and which, like winter snow, can be seen in January.

It would be wrong, however, to assume that all of the meteor showers recorded by Homer are exactly the same ones that can be seen today. Because the Earth passes through various belts of meteorite dust on its passage through space, meteor showers seen today may not all have been visible in Homer's time, and conversely there may have been meteor showers then which are not visible now.

When Odysseus takes off his armour (four stars in the torso of Boötes), or flings his cloak to the ground, Homer uses the discarded equipment to draw attention to stars outside the body of the constellation, as in the following quotations. 'His

[Odysseus'] armour is laid upon the ground, and he stalks in front of the ranks as it were some great woolly ram ordering his ewes' (3.195). 'He [Odysseus] flung his cloak from him and set off to run. His servant Eurybates, a man of Ithaca, who waited on him, took charge of the cloak' (2.183). Odysseus' armour and clothing are stars near the 'feet' of the constellation, the clothing on one side and the armour on the other (figs. 40 and 41). The clothing is placed near the left foot of Odysseus/Boötes because there it is closer to the personal star of Eurybates, who picks up the garments.

Two other quotations also relate Odysseus' armour to Boötes. 'On his head he set a leathern helmet that was lined with a strong plaiting of leathern thongs, while on the outside it was thickly studded with boar's teeth, well and skilfully set into it; next the head there was an inner lining of felt' (10.261). Odysseus' famous boar's-tooth helmet has a long history in myth, its description being passed down from generation to generation. The upper part of his helmet is identifiable as a shallow dome of faint stars in the head of Boötes. 'On this Odysseus went at once into his tent, [and] put his shield about his shoulders' (10.149). The stars of Corona Borealis make a fine shield, and their relationship to Boötes gives them the impression of being carried on the shoulder of Odysseus.

Diomedes as Perseus

The constellation of Perseus has been depicted as a warrior since time immemorial, so it is not surprising that in the *Iliad* it is represented by Diomedes, one of the most powerful fighters in the Greek army. His deeds of derring-do and Homer's descriptions of his body and clothing give a graphic picture of Perseus (fig. 42), with particular emphasis on the long legs and feet and the shoulders, arms, stomach and heart of the constellation.

Without astronomical instruments it is probably impossible to determine whether or not stars at the foot of Perseus lie just

Fig. 42 *Perseus as the Greek warrior Diomedes.*

inside the zodiac, home of the gods that in the *Iliad* represent
the Moon and the five planets visible to the naked eye. If the
stars *were* in the zodiac, the epithet 'god-like' could have been
given to the constellation. Homer raises this problem in ingenious fashion, but his conclusions are not decisive. He puts his
opinion into the words of Glaucus (6.128) and of Pandarus,
who says he cannot tell whether Diomedes (Perseus) is a god
or not, but even if he is not he 'has some god by his side who
is shrouded in a cloud of darkness' (5.185). This refers to
Athene, the planet Jupiter (see fig. 43).

In the same manner and for the same reasons that
Agamemnon berates Odysseus during his tour of inspection,
the Greek commander-in-chief also admonishes Diomedes for
not being in the front line of the action. 'Son of Tydeus,' says
Agamemnon, 'why stand you cowering here upon the brink of
battle?' (4.371). Again, this is a fine use of narrative to point out
that warriors in Triangulum and Pegasus, the constellations

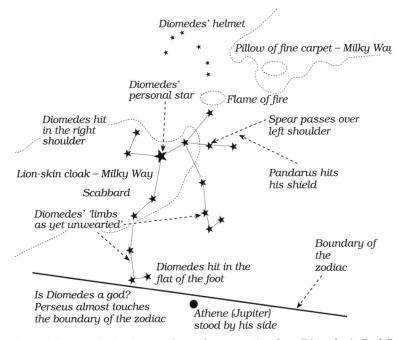

Fig. 43 *The constellation Perseus drawn from narrative about Diomedes in Book 5 of the* Iliad.

that precede Perseus through the sky, play no great roles in the fighting. Mirphak, α Persei, the personal star of Diomedes, is located at the centre of the constellation when Athene says, 'Fear not, Diomedes, to do battle with the Trojans, for I have set in your heart the spirit of your knightly father Tydeus' (5.125).

Ancient records of meteor showers are sparse, and the earliest record of the Perseids dates back to AD 36. Homer knew of such events much earlier, and says these 'flames' came from Diomedes' shield and helmet and his head and shoulders: 'She [Athene] made a stream of fire flare from his shield and helmet like the star that shines most brilliantly in summer after its bath in the waters of Oceanus – even such a fire did she kindle upon his head and shoulders as she bade him speed into the thickest hurly-burly of the fight' (5.4). Nowadays, the radiant (or apparent source) of the Perseids meteor shower moves across the north of Perseus during a twelve-day period from early to mid August, and the shower is at its peak when in the region of Diomedes' shining helmet. (See also Chapter 6, page 184.) There is another, less conspicuous, shower of very swift meteors from 25 July to 4 August in the region of α–β Perseus,[4] and Diomedes is wounded in that area and sheds drops of blood: '[Pandarus] aimed an arrow and hit the front part of his cuirass near the shoulder: the arrow went right through the metal and pierced the flesh, so that the cuirass was covered with blood' (5.98). '"Dear son of Capaneus," said he [Diomedes], "come down from your chariot, and draw the arrow out of my shoulder." Sthenelus sprang from his chariot, and drew the arrow from the wound, whereon the blood came spouting out through the hole that had been made in his shirt' (5.109).

More than once Paris of Troy fires his arrows across the heavens, either to establish a 180° panorama of the skies or to mark a straight 'line of sight' between stars in two constellations (see Chapter 4). A line drawn from Paris (α Orionis) extends to ζ Persei, the flat foot of Perseus: 'Paris drew his bow and let fly an arrow that sped not from his hand in vain, but pierced the flat of Diomedes' right foot, going right through it and fixing itself in the ground' (11.375).

As with other heroes, the clothing, armour and weapons of Diomedes identify stars that can be seen in figs. 42 and 43. Quotations describing Diomedes in this manner include: 'On his head [Thrasymedes] set a helmet of bull's hide without either peak or crest; it is called a skull-cap and is a common headgear' (10.257). '[Pandarus] poised his spear as he spoke and hurled it from him. It struck the shield of the son of Tydeus [Diomedes]; the bronze point pierced it and passed on till it reached the breastplate' (5.280). In the funeral games, 'Aias pierced Diomedes' round shield, but did not draw blood, for the cuirass beneath the shield protected him' (23.817). 'Thrasymedes provided the son of Tydeus [Diomedes] with a sword and a shield for he had left his own at his ship' (10.255). The sword and shield of Diomedes are powerful images; the shield is created from the stars of Perseus, and the sword represents the Milky Way and is used to kill the spy Dolon and to dispatch the Thracians during a night raid on the Trojan lines.

In the *Iliad* more Trojan warriors are killed than Greeks, and this means that fewer stars in 'Greek' constellations are identified during the fighting. To overcome this problem, Homer used the funeral games in Book 23 to describe in more detail the principal Greek constellations. In some cases contestants win prizes – or stars – they already own. Meriones, for instance, is awarded two talents of gold, which represent his personal double star, θ^1 and θ^2 Tauri (23.614). Diomedes is another contestant who also wins a prize he already owns – the large silver sword and scabbard he is awarded are already part of his constellation (23.824).

The Milky Way runs prominently through Perseus, and it becomes for Diomedes the lion skin he wears, and the fine carpet on which he rests his head while sleeping on the skin of an ox: 'Diomedes threw the skin of a great tawny lion about his shoulders . . . that reached his feet' (10.177). 'The hero was sleeping upon the skin of an ox, with a piece of fine carpet under his head' (10.156).

Achilles as Canis Major

In this short survey of warriors as constellations, Achilles, whose personal star is Sirius, α Canis Majoris, has been left to last, because he is probably the most complicated and references to him in the *Iliad* are numerous. The allocation of the most powerful warrior at Troy to the brightest star in the sky is obvious, but it is when Achilles is seen as the entire constellation (fig. 44) that matters become more interesting.

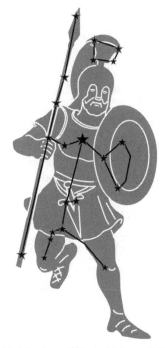

Fig. 44 *The stars of Canis Major and Achilles.*

When we first searched the *Iliad* for references to support the image of Achilles as a constellation, it was known (see Chapter 7) that the astronomical history of the region of the sky centring on Canis Major was of major importance. Homer did not divide the skies into as many constellations as there are today, and so the boundaries of at least some of them would have been larger. Since ancient times, charts of the heavens have periodically been revised and new constellations have been added either from the stars of older constellations or from stars in the southern hemisphere. Homer defines forty-five constellations in the Catalogue of Ships, but today eighty-eight are defined.

For reasons explored in Chapter 7, we believe that the constellation home of Achilles included Canis Major, Monoceros and Canis Minor, as well as the star β Cancri. Fig. 45 illustrates

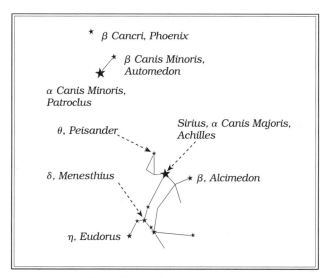

Fig. 45 *Canis Major, with Achilles and the commanders of the Myrmidons representing brighter stars. Patroclus and Automedon are in Canis Minor, while Phoenix belongs to Cancer.*

how the brighter stars of this wide area of sky are allocated to Achilles, his friends Patroclus, Automedon and Phoenix, and the other three captains of the regiment of the Myrmidons, Menesthius, Eudorus, and Peisander. When Phoenix, Aias, Odysseus and two heralds are sent from the 'hut' of Agamemnon (9.178) to the huts or tents of the Myrmidons to try to persuade Achilles to return to the battle, they find Achilles and Patroclus in the same 'hut' (9.190), the ancient boundary for the stars of his regiment. Greek leaders such as Idomeneus and Meriones (Taurus) and Eurypylus (Lyra) also have 'huts' or 'tents' – the areas of sky encompassed by their respective constellation boundaries. Achilles is in his celestial 'hut' when he is 'sitting opposite' his friend Patroclus; it can be seen from fig. 45 that this places their personal stars, Sirius, α Canis Majoris, and Procyon, α Canis Minoris, at opposite ends of the constellation boundary defined by Homer.

Traditionally Canis Major has been seen as a dog, but Homer uses the same stars to give two different manifestations of the usual configuration of the stars of Canis Major. When Achilles is seen as the entire constellation in human form, the brighter stars are highlighted by descriptions of his heart (Sirius, the brightest star), his chest, liver, forearm, hands, knees and feet to give the impression of a man. Stars defining his shield, helmet, spear and sword add the accoutrements of war to this outline. In Book 18, Achilles' mother, Thetis, suggests yet another image of the significantly brighter stars of the constellation when she describes her son as a young sapling who will grow into a mighty warrior. As is shown in fig. 69, Homer also uses Canis Major to reiterate his idea of the celestial sphere when, in Book 23 of the *Iliad*, Achilles tosses this way and that during a restless sleep.

ACHILLES THE WARRIOR

The brighter stars of the ancient and conventional view of Canis Major as a dog, together with the stars of Canis Major and Monoceros, are used to identify Achilles (Sirius), his friends Phoenix and Patroclus, and the captains of the regiment of Myrmidons (fig. 45). Homer's description of Canis Major reaches imaginative heights when he describes how the constellation was created to accommodate the return to the skies of the star Sirius, and how Hephaestus, the smith-god, forges new armour for Achilles – see Chapter 7. From Achilles' great spear to his tight-fitting armour and glittering helmet, Homer lavishes literary description on the most powerful warrior at Troy, and, by implication, Canis Major.

In fig. 46 the stars of Canis Major are used as the body of Achilles. His personal star, Sirius, lies at the heart of the constellation, whether it is viewed as a Great Dog or as a Greek warrior. Other physical attributes – his throat, arms, hands, liver, thighs, knees and shin, together with his weapons and armour – contributing to the creation of a warrior in the

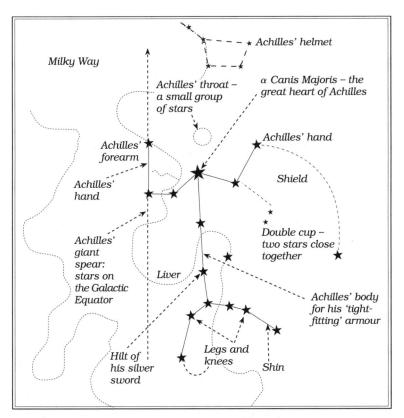

Fig. 46 While largely retaining the shape of the traditional Canis Major, this diagram shows how the same stars match narrative associated with Achilles.

heavens are found in the following quotations. 'Antilochus bent over him [Achilles] the while, weeping and holding both his hands as he lay groaning for he feared he might plunge a knife into his own throat' (18.34). 'I will go up against him though his [Achilles'] hands . . . be fire and his strength iron' (20.371). 'With the other spear he [Asteropaeus] grazed the elbow of Achilles' right arm drawing dark blood, but the spear itself went by him and fixed itself in the ground, foiled of its bloody banquet' (21.167). '. . . that terrible man on whose liver I [Hecuba] would fain fasten and devour it' (24.213). 'Whereon

Achilles smote his two thighs' (16.125). 'But Lycaon came up to him dazed and trying hard to embrace his knees, for he would fain live, not die' (21.74). 'And the spear struck Achilles on the leg beneath the knee' (21.592). 'There was also Tros, the son of Alastor – he came up to Achilles and clasped his knees in the hope that he would spare him and not kill him but let him go, because they were both of the same age' (20.464). The implication behind this last remark is that the personal star of Tros (τ Cetii) is close to the same altitude in the sky as Sirius, and would have appeared on the horizon visible from Greece at about the time Sirius also appeared in sky, *c*. 8900 BC.

Descriptions of Achilles' armour and effects that match the qualities of Canis Major include the following: 'He [Achilles] stayed his hand on the silver hilt of his sword, and thrust it back into the scabbard' (1.190). 'He slung the silver-studded sword of bronze about his shoulders, and then took up the shield so great and strong that shone afar with a splendour as of the moon' (19.374). 'Asteropaeus failed with both his spears, for he could use both hands alike; with the one spear he struck Achilles' shield' (21.165). 'His [Hector's] aim was true for he hit the middle of Achilles' shield, but his spear rebounded off it' (22.290). 'He lifted the redoubtable helmet, and set it upon his head, from whence it shone like a star, and the golden plumes which Hephaestus had set thick about the ridge of the helmet, waved all around it' (19.380) – Achilles' flashing helmet is in the Milky Way, and the plumes are stars of Monoceros. 'All night long did Achilles grasp his double cup' (23.218). 'Double cup' is a phrase used elsewhere in the *Iliad* for visible pairs of stars, such as ξ^1 and ξ^2 Canis Majoris.

Chapter 7 examines Homer's theory of the precession of the equinoxes and explains the astronomical significance of the shield made for Achilles by the crippled smith-god, Hephaestus. The shield and armour are allegories for the reorganization that became necessary when Sirius reappeared in the skies of Greece.

ACHILLES AS A SAPLING

'I [Thetis] bore him fair and strong, hero among heroes, and he shot up as a sapling; I tended him as a plant in a goodly garden' (18.438). This variant image of Achilles as a young tree helps fix the arrangement of Canis Major standing on its 'tail' (fig. 47).

The Catalogue of Ships and the Constellations

To list all the stars and constellations recorded by Homer is beyond the scope of this introduction to his astronomy, but a short directory based on the Catalogue of Ships in Book 2 will expand on the examples given in this chapter. Tables 7 and 8 match the place of origin of each of the forty-five Greek and Trojan regiments to the modern name of a constellation and identify the warriors associated with principal stars in the constellation. For example, Mycenae represents the constellation Leo, of which the star α Leonis is represented by Agamemnon. Regiments are listed in the order in which they appear in the Catalogue of Ships. (See also Chapter 8.)

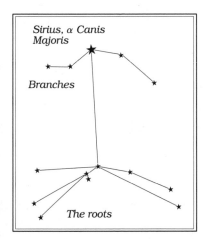

Fig. 47 *Achilles the sapling.*

Table 7 *Greek Regiments, their constellations and commanders*

Place of origin	Commanders
1. Boeotia = Draco	η Peneleos
	λ Leitus
	α Arcesilaus
	ι Prothoenor
	κ Clonius
2. Aspledon and Minyaean Orchomenus = Pleiades (Taurus)	η Ascalaphus
	27 Ialmenus
3. Phocis = Cygnus	α Schedius
	β Epistrophus
4. Locris = Cepheus	α Little Aias
5. Euboea = Cassiopeia	γ Elephenor
6. Athens = Crux Australis	α Menestheus
7. Salamis = Argo	α (Carinae) Great Aias
8. Argos = Perseus	α Diomedes
	β Sthenelus
	γ Euryalus
9. Mycenae = Leo	α Agamemnon
10. Lacedaemon = Scorpius	α Menelaus
11. Pylos = Auriga	α Nestor
12. Arcadia = Leo	ι Agapenor
13. Buprasion and Elis = Cancer	γ Amphimachus
	ι Thalpius
	α Diores
	δ Polyxenus
14. Dulichium and Echinean Islands = Canes Venatici	α² Meges
15. Cephallonia = Boötes	α Odysseus
16. Aetolia = Aries	α Thoas
17. Crete = Taurus	α Idomeneus
	θ Meriones
18. Rhodes = Pisces	η Tlepolemus
19. Syme = Triangulum	α Nireus
20. Calydnian Isles = Andromeda	α Antiphus
	β Pheidippus
21. Pelasgian Argos (Phthia) = Canis Major	α Achilles
22. Phylace = Ursa Minor	α Protesilaus
	β Podarces
23. Pherae = Pegasus	α Eumelus
24. Methone = Hercules	α Philoctetes
	τ Medon
25. Tricce = Ophiuchus	α Podalirius
	β Machaon
26. Ormenion = Lyra	α Eurypylus
27. Argissa = Aquila	α Polypoetes
	γ Leonteus
28. Cyphus and Dodona (Enienes ... and Peraebians) = Sagitta	γ Guneus
29. Magnetes = Delphinus	α Prothous

Although Patroclus is a leading warrior in the *Iliad*, he is not named in the Catalogue of Ships. His position in the skies as α Canis Minoris is indicated by narrative

Table 8 Trojan regiments and their commanders

Commander or place of origin	Commander
1. Hector = Orion	β Hector
2. Aeneas = Virgo	α Aeneas
	β Archelocus
	δ Acamas
3. Zelia = Sagittarius	σ Pandarus
4. Adresteia = Equuleus	δ Adrestus
	γ Amphius
5. Percote = Aquarius	α Asius
6. Pelasgian Larissa	η Hippothous
= Ophiuchus	θ Pylaeus
7. Thrace = Capricornus	β Acamas
	δ Peiros
8. Ciconia = Aquarius	ε Euphemus
9. Paeonia = Centaurus	θ Pyraechmes
10. Paphlagonia = Lupus	γ Pylaemenes
11. Halizoni = Crater	δ Odius
	α Epistrophus
12. Mysia = left foot of	η Chromis
front twin (Gemini)	μ Ennomus
13. Phrygia = Corvus	γ Phorcys
	β Ascanius
14. Meonia = Pisces	α Mesthles (Grus)
Australia and Grus	β Antiphus (Grus)
15. Caria = α and	α Amphimachus
β Centauri	β Nastes
16. Lycia = Gemini	β Sarpedon
	α Glasucus

King Priam, leader of the Trojans, takes no part in the fighting and is placed at ζ Ursae Majoris.

6

Gods in the Heavens

Would that I [Hector] were as sure of being
immortal and never growing old, and of
being worshipped like Athene and Apollo.
Iliad 8.538

Throughout history it has been common to look towards the
heavens for a solution to the mystery of the creation of the
universe and the place of mankind within it. The origins of
Homeric astronomy lie deep in the oldest Greek creation
myths – stories of gods and supernatural events conceived to
ease the human need to understand and to explain. Both
benign and destructive natural forces – from the Sun and the
rains to storms and tempests – emanate from above, and,
together with the glorious panorama of the night skies, inspire
awe, fear and curiosity. In ages long past, the apparent vastness
of space and the passage of the Sun, the Moon, the stars and
the five planets visible to the naked eye raised questions that
could not be explained by human experience.

It is from *Theogony*, or *Birth of the Gods*, by Homer's contemporary Hesiod that we learn about the earliest Greek gods and
supernatural myths of creation. From primeval chaos or the
Abyss came Gaia (the Earth), Tartarus, Eros, Erebos and Night.
Earth's first child was Ouranos, or the starry heaven, and she

also gave birth to such strange creatures as the gigantic and brutal Titans, the one-eyed Cyclopes, and three monster children each with fifty heads and a hundred arms. Hesiod's verse tells of the creation:

> From Chaos came black Night and Erebos
> And Night in turn gave birth to Day and Space
> Whom she conceived in love to Erebos.
> And Earth bore starry Heaven, first, to be
> An equal to herself, to cover her
> All over, and to be a resting-place,
> Always secure, for all the blessed gods.[1]

The climax to this fantasy is the story of Cronus, the 'crooked-scheming' Titan who had married his own sister, castrated his father, and then deposed him as ruler of the world. Fearing that he in turn would be overthrown by his own children, Cronus swallowed his first five children – Poseidon, Hades, Hera, Demeter and Hestia – as they were born, but a sixth, the baby Zeus, was spirited away to a cave on Crete, from where, in his maturity, he led a rebellion. Cronus' consort forced him to vomit up his first five children still alive, and they joined Zeus in a long and successful war against their father. As the last leader of the first order of gods, Cronus had served his purpose in explaining the mythical creation of the cosmos, and he was dispatched to languish for ever in the deepest realms of the underworld.

Myth took on a more practical nature in at least one of its facets when it became a vehicle to preserve learning about what can be seen in the skies by the human eye. In Homer's *Iliad*, the activities of the Olympians, led by Zeus, are allegories for celestial phenomena, and record the nature of the universe and the movement of the heavens. The Siege of Troy may be a war between mortals, but almost half of the *Iliad* is occupied with the quarrelling of the gods and their interference in the affairs of man. It is not suggested that the pantheon of gods who, with their associated religious rituals, thrived after the demise of Cronus had no function in Homeric society other than to act as astronomical icons. Nor is it likely that Homer merely hijacked the gods from their mythic and religious environments to suit the astronomical purpose of the *Iliad*. Rather, myth and religion – two powerful elements in the structure of ancient societies – were intertwined with the age-old practical but also mysterious science of astronomy.

Given the large numbers of ancient deities named by Hesiod or honoured in later Greek times, Homer is quite economical in his use of the gods in *active* roles. His major players – the brothers, sister/consort and children of Zeus – are just sufficient in number to accommodate significant astronomical ideas such as the celestial sphere, or objects such as the Moon and the five planets visible to the naked eye that are known in modern times as Saturn, Jupiter, Mars, Venus and Mercury. Lesser gods represent the Milky Way and the horizon.

It will never be known when astronomical learning was first woven into stories about gods and mortals, but it was certainly long before Homer. It is reasonable to assume that it was a cumulative process, and evidence from the *Iliad* suggests that new stories and mortal characters were introduced as astronomical knowledge expanded. An indication of the possible age of elements of these myths has already been noted in the recording in the *Iliad* of astronomical events such as the return to the skies of Sirius, which began in the ninth millennium. A repetitive feature of the narrative is reminiscence of earlier ages when an older generation of heroes ruled the earth. King Nestor in particular rambles on about his youth (e.g. at 1.260) – a time when Hercules rampaged through the land (11.669), before dying and being placed in the skies as a constellation. It may be no coincidence that the star that indicated celestial north as long ago as *c.* 7000 BC was τ Herculis, and its predecessor during the previous two thousand years or more was ι Herculis.

The Olympian gods who to Homer would have been old and familiar acquaintances reflected influences from much earlier times, and even from beyond the borders of Greece. 'Modern scholars believe that the religious system presented by Homer and Hesiod arose gradually, starting with a fusion of two or three different cultures after 2100 BC. But the facts are lost in prehistory, and scholars must rely on scant evidence from archaeology in order to trace this process of religious fusion.'[2]

Homer has Poseidon describe the apportioning of the universe between him and his brothers Zeus and Hades, in this manner:

> We were three brothers whom Rhea bore to Cronus – Zeus, myself, and Hades who rules the world below. Heaven and Earth were divided into three parts, and each of us was to have an equal share. When we cast lots, it fell to me to have my dwelling in the sea for evermore; Hades took the darkness of the realms under the Earth, while air and sky and clouds were the portion that fell to Zeus; but Earth and great Olympus are the common property of all. (15.187)

In outlining this three-way division of the spoils of inheritance after the fall of Cronus, Homer is declaring his idea of the universe being contained within a celestial sphere (fig. 48). Interpretation of further allegorical narrative in Chapter 8 will show that his view was strikingly similar to the one that survived until the sixteenth century AD.

As well as being the supreme god, Zeus' specific area of influence was the panorama of sky from east to west and north to south that is visible to an observer on Earth. Hades controls the underworld, that part of the celestial sphere below the horizon, and it is significant that he never appears on the battlefield of Troy – the visible skies. The role of Poseidon is not so straightforward, and it may eventually be shown to have more than a

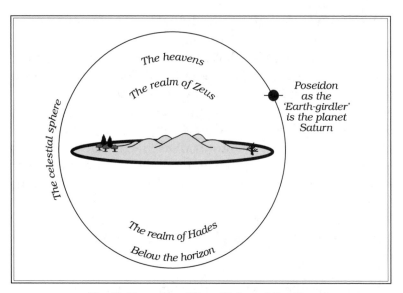

Fig. 48 *The realms of Zeus, Hades and Poseidon.*

single strand, but it is the epithets such as 'Earth-encircler' (Butler translation) or 'Earth-girdler' (E. V. Rieu translation) commonly applied to him that concern us now. He takes part in similar astronomical activities to gods who are more readily associated with planets. These include Poseidon 'covering' or assuming the form of mortals (11.749, 13.44 and 14.135) – events that indicate the occultation or temporary obscuring of stars by a planet. Other phrases create images of the rising (20.15) and setting (15.218) of Saturn, and of its journey across the heavens in company – or in conjunction – with other planets (20.115, 20.132, 21.435 and 21.462). For these reasons Poseidon has been assigned to the planet known in later history as Saturn.

Poseidon's role as god of the seas and earthquakes suggests that he had physical connotations beyond those of astronomy, and these perhaps accrued to him in ancient times as part of a view of the universe that included not only the skies but also the Earth and seas.

An indication that gods and goddesses are associated with planets is contained in the papers of Edna Leigh, but her notes are sparse. It is not, however, difficult to link beautiful Aphrodite with Venus, the fairest of all the planets, nor to match Ares, the fiery god of war, with Mars, the red planet long associated with war and bloodshed. In post-Homeric Greece the gods of love and war became identified with the planets Venus and Mars. The astronomical roles of two other principal goddesses, Hera and Athene, were not so easily assigned, but an examination of the Judgement of Paris, the most famous beauty contest in history, confirmed Aphrodite as Venus and led us to believe that Hera represents the Moon and Athene Jupiter. An extensive search of the *Iliad* for every reference to Hera and Athene supported these identifications.

The three goddesses from whom Paris was asked to choose the most beautiful would seem to represent the most beautiful three objects in the night sky: Venus, the Moon and Jupiter. The glittering planet Venus – the third brightest object in the sky after the Sun and Moon, and beautiful without compare in the night sky – must represent Paris's eventual choice, Aphrodite. Hera is brightest of the three and beautiful at the time of the crescent and full Moon, but over the course of a lunar month as her illuminated 'face' waxes and wanes it becomes deformed in its gibbous phase between half and full moon. Athene was allocated the role of Jupiter, the steadfast shining planet that is second only to Venus in brightness. With their possible identities derived from the Judgement, we examined the actions of Hera and Athene in the *Iliad* for associations with singular attributes of the Moon and Jupiter. As we shall show, the results were pleasing.

In a number of cases when two or even three gods were involved in a specific event, they could be associated with astronomical phenomena jointly concerning their respective planets. From these and other inquiries we propose that the major gods and planets should be aligned in the following manner:

Hera = the Moon
Aphrodite = Venus
Athene = Jupiter
Ares = Mars
Apollo = Mercury at dawn
Hermes = Mercury at dusk
Poseidon = Saturn

Endorsement of the view that the planets visible to the naked eye were represented by gods came about in an unusual way. After two years cataloguing the Leigh papers, we began, warily at first, to derive astronomical images from the narrative. One of the first came from a few words in Book 1 that record the return of the gods after a few days' feasting in Ethiopia: 'The immortal gods came back in a body [at dawn] to Olympus, and Zeus led the way' (1.493). An image was created of Homer describing a time when the crescent Moon and the five planets visible with the naked eye could be seen close together in the sky at dawn and rising towards the zenith of the heavens, the domain of Zeus. Our inquiries into whether such a conjunction of planets is rare or common were not at first successful, but eventually the happy chance of reading an article in an astronomy magazine[3] showed that such a conjunction on 5 March 1953 BC had been recorded in an ancient Chinese text and had so impressed Chinese astronomers that it was used as a new starting point for their calendar. That conjunction would also have been visible from Greece. In fact researchers Kevin D. Pang and John A. Bangert discovered that the five naked-eye planets had been in the sky for some days before that date, and on 26 February 1953 BC their alignment had been closer than at any time in the past six millennia. But extra significance was given to the conjunction

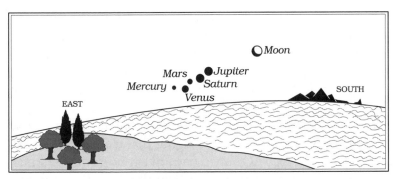

Fig. 49 Dawn on 5 March 1953 BC.

when the new moon joined the planetary grouping at dawn on 5 March (fig. 49), thus adding the goddess Hera to the planets represented by Athene, Aphrodite, Poseidon, Ares and Apollo.

Although they differ in literary expression, a number of translations of this event from the *Iliad* reflect similar astronomical images. Butler's translation quoted above is quite austere, but others specifically make reference to the dawn, and state that the gods returned in 'a line' (E. V. Rieu), 'a cortège' (Robert Fagles) or 'are for ever, all in one company' (A. T. Murray). A literal translation from the Greek texts says 'from that [time] became the twelfth dawn, then indeed to Olympus went all the ever-being gods, all at the same time, and Zeus led'.[4]

Whether the conjunction described by Homer is the same one that so influenced the Chinese may never be known, but so rare are such events that it is possible that it was, and had remained alive in popular memory in Greece through being preserved in allegory.

In the course of the *Iliad*, gods (representing planets) assume the voices or characters of mortals (stars). In this way Homer records a phenomenon known as occultation, when the Moon or a planet journeying across the sky passes in front of celestial

163

objects that are further away, such as stars or outer planets, and temporarily obscures them. It may be that total obscuration occurs only when a god assumes the *body* of a mortal; when a god assumes the *voice* of a mortal or is said to be *helping* him, this may indicate that the planet is only close to the personal star of the warrior concerned.

The Moon, the planets and the divisions of the universe account for only nine of the active gods that Homer includes in the *Iliad*, and we also examined narrative concerning Oceanus, Iris and Xanthus, together with that about Hephaestus, the great artificer, for astronomical significance.

Oceanus is proposed as the god who rules the river of the same name that was said to encircle the Earth and into which the Sun set each night. Experience of those who had sailed the wide seas beyond the horizon visible from land, and had never come across such a river, must raise doubts about the Greeks taking this as a literal concept. If, however, Oceanus is designated as god of the horizon, then another part of Homer's universe falls into place. In hot climates, the line where the skies meet the Earth can be disturbed by dust and atmospheric turbulence and may appear to be a shimmering band, perhaps even like sunlight reflected from a river.

Iris – in later times to be honoured as the goddess of the rainbow that arches across the sky – carries messages from the heights of Olympus (the heavens) to the Earth or into the sea. At 2.786 Homer says 'Iris, fleet as the wind, was sent by Zeus to tell the bad news among the Trojans' on Earth, and at 15.172 she 'did she wing her way till she came close up to the great shaker of the Earth [Poseidon]'. In this messenger role she is thought to be associated with the concept of the dome of the heavens and the arc of the celestial sphere.

The god Xanthus is identified as one of Homer's manifestations of the Milky Way, the heavenly river of light. Homer describes the Milky Way which lies alongside the constellation of Canis Major when Achilles forces Trojans to flee in front of him until their path was blocked by 'the deep silver-eddying stream, and [they] fell into it with a great uproar . . . As locusts flying to a river before the blast of a grass fire – the flame comes on and on till at last it overtakes them and they huddle into the water – even so was the eddying stream of Xanthus filled with the uproar of men and horses' (21.8).

Hephaestus, the great artificer, was the builder of the mansions of the gods, and in the *Iliad* he creates the glittering new armour and shield for Achilles' return to the battlefield. As we shall see, narrative about Hephaestus is quite specific in defining his astronomical role, but the image it creates leads to a field that extended the roles of the gods and had hitherto been perplexing.

Insofar as Homer gives some mortals dual roles as stars and constellations (see Chapters 4 and 5), it is perhaps not surprising that major gods may also represent constellations as well as planets. Hephaestus is one of these and, as well as being the artificer who created the constellations, he has as his personal abode in the skies the constellation of Perseus, which he shares with Diomedes.

Creating an astronomical portrait of a particular god is a painstaking process that begins with the extraction of every reference to the deity from the pages of the *Iliad*. In the case of Apollo, for instance, there are more than eighty occasions on which he is active, while for Athene there are more than a hundred. These extracts – ranging from a few words to several paragraphs – are examined for astronomical content, and as detail is added to detail a mental image of the god and his or her astronomical purpose becomes clearer. There follow profiles of Hera as the Moon, Ares as Mars, Athene as Jupiter, Aphrodite as Venus, Apollo as Mercury at dawn, Hermes as Mercury at dusk and Hephaestus as Perseus. The chapter concludes with an analysis of Thetis and her nymphs as the constellation of Eridanus.

165

Hera as the Moon

Hera, queen of Heaven and the sister and consort of Zeus, is also the goddess of marriage and the protector of women in childbirth. This latter role fits well her association with the Moon, the celestial object that is reborn each lunar month. Hera is as important in astronomical allegory as she is in the narrative of the *Iliad*. A poor loser in the Judgement of Paris, she displays her anger and hatred of the Trojans from the opening pages of the epic. Her title of queen indicates an ancient perception of the Moon as queen of the night skies, and references to her as 'strange Queen' (1.560 and 4.31, A. T. Murray translation) suggest the distorted or unbalanced face of the Moon in its gibbous phase between half and full Moon.

Hera is ingeniously identified with the Moon when Homer uses favourite epithets to describe her throughout the epic. During the course of Book 1, she is first said to be 'white-armed'[5] and then 'ox-eyed', before becoming again 'white-armed' – rather meaningless phrases unless they are considered astronomically. Similar instances, occurring throughout the *Iliad*, describe a monthly lunar cycle, from the waxing crescent of the new Moon to the full Moon and then on to the waning crescent of the old Moon:

White-armed = new crescent Moon = ☽
Ox-eyed = full Moon = ○
White-armed = old crescent Moon = ☾

During the seduction scene in Book 14, Hera's usual aggressive mood changes. Homer uses sublime language to describe her captivation of Zeus, which is in astronomical terms a fine portrayal of a spectacular lunar eclipse. In depicting a beautiful woman preparing for a romantic encounter, Homer's words draw an image of the full Moon in all its beauty, describing its 'seas', the halo or veil with which it is sometimes surrounded, and its soft light spreading like perfume over the Earth and skies:

> [Hera] cleansed all the dirt from her fair body with ambrosia, then she anointed herself with olive oil, ambrosial, very soft, and scented specially for herself – if it were so much as shaken in the bronze-floored house of Zeus, the scent pervaded the universe of Heaven and Earth. With this she anointed her delicate skin, and then she plaited the fair ambrosial locks that flowed in a stream of golden tresses from her immortal head. She put on the wondrous robe which Athene had worked for her with consummate art, and had embroidered with manifold devices; she fastened it about her bosom with golden clasps, and she girded herself with a girdle that had a hundred tassels: then she fastened her earrings, three brilliant pendants that glistened most beautifully, through the pierced lobes of her ears, and threw a lovely new veil over her head. (14.170)

The 'seas' of the Moon are physical features that stand out sharply on a clear night. One of Hera's earrings with its triple pendants is a metaphor for three 'seas' on the western side of the Moon when viewed from Earth: Mare Crisium, Mare Fecunditatis and Mare Nectaris.[6] On the eastern side, Mare Imbrium, Mare Humorum and Mare Nubium make up the matching earring. In Leaf's translation the earrings are described as 'earrings of triple drops', but a literal translation of the phrase is said to imply mulberries, and Robert Fagles writes of 'ripe mulberry clusters dangling in triple drops' – another striking image for the 'seas'. Hera's fair ambrosial locks and lovely new veil are the halo which sometimes surrounds the disc of the full Moon, and the scent that spreads over the Earth and skies is the Moon's gentle light.

With Hera dressed in her finery, she works her magic on Zeus, and the moment of seduction describes a total lunar eclipse (Table 9, *overleaf*). While Homer draws this image of a total lunar eclipse, he does not attempt to describe the astronomical reasons for it. This suggests that he was unaware that a lunar eclipse occurs when the Moon passes through the Earth's shadow.

Table 9 Hera's seduction scene and the Moon

Literary narrative of the seduction scene	Seduction scene as a lunar eclipse
1. The sequence opens with Hera being 'ox-eyed'.	1. A lunar eclipse can occur only at full Moon.
2. Aphrodite leaves the scene.	2. As Venus, the 'evening star', sets two hours or so after dusk, Homer establishes that it is night-time.
3. Hera leaves her 'room' and travels from Olympus to Mount Ida.	3. The Moon is in the night sky.
4. Hera discusses her plan with 'Sleep', both of whom are then 'covered in a mist'.	4. In the first stages of an eclipse the Moon gradually moves into the penumbra (the outer part of the shadow of the Earth) and its face becomes a little less bright.
5. Hera is now described as 'white-armed'.	5. As the Moon moves into the umbra (the inner part of the shadow), a crescent of bright light is the last part seen of the Moon before it is in total shadow.
6. Hera is afraid their lovemaking will be visible to all, but Zeus says he will hide her in a golden cloud through which even the Sun's rays cannot penetrate.	6. The eclipse will be total.
7. As they lie down, they are covered by a golden cloud from which fall drewdrops.	7. The eclipse is total but the Moon is bathed in Earth-shine – sunlight refracted by the Earth's atmosphere faintly illuminates the Moon with a coppery light.
8. Zeus holds Hera in his arms, and they are overcome by love and sleep.	8. The Moon is completely hidden.
9. Zeus awakes.	9. The lunar eclipse is over.

Homer draws attention to the physical appearance of the Moon in the guise of Hera when on several occasions he associates it with human feelings and a popular, even childlike, image of its face. When all of Hera's schemes are going well in the closing lines of Book 1 she is said to have a 'smiling face' (1.596), but when thwarted in Book 8 she shakes with rage (8.198). Confronted with Zeus' anger in Book 15, she is forced to control her feelings, and signs of both happiness and bad mood are combined: 'On this Hera sat down, and the gods were troubled throughout the house of Zeus. Laughter sat on her lips but her brow was furrowed with care, and she spoke up in a rage' (15.100).

We believe Homer was recording an unusual astronomical phenomenon when he has Zeus sternly remind Hera, 'Do you not remember how once upon a time I had you hanged? I fastened two anvils on to your feet, and bound your hands in a chain of gold which none might break, and you hung in mid-air among the clouds' (15.18). These lines suggest a lunar phenomenon when four other 'moons' can be seen in the skies. Known as 'parselene', 'mock moons' or 'moon dogs', these are small diffuse images that can be seen at 22° on either side and north and south of the Moon. They are caused by the refraction of light by ice crystals in the atmosphere, and, although it is possible to see several moons, it is most common to see only one or two at a time, like the anvils tied to Hera's feet.

Hera is also associated with astronomical events such as conjunctions, when two or more heavenly bodies appear to be close together in the sky. This is an illusion caused by bodies such as a planet and the Moon, widely separated in space, being in alignment when seen from the Earth. Hera's role in such occurrences can be seen in the following profiles of Ares and Athene.

Ares as Mars

Ares, the god of war, in the form of the planet Mars, is one of the most prominent and interesting objects in the sky. As the fiery red planet journeys across the heavens, it evokes the image of a peppery colonel of hussars rushing hither and thither across the battlefield

In addition to its striking colour, another notable characteristic of Mars is the manner in which, like other outer planets, it appears to reverse its journey across the sky and travel

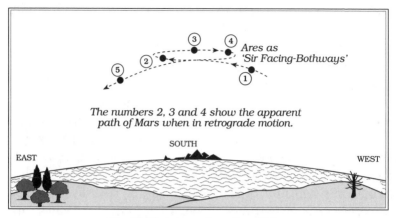

Fig. 50 *The retrograde motion of Mars.*

backwards for several weeks, before resuming its usual path (fig. 50). Known as retrograde motion, this is an optical illusion caused by the differing orbits of Earth and Mars around the Sun.

Ares plays a prominent role in Book 5 of the *Iliad*, and when the narrative is examined for astronomical content it draws a remarkable picture of the motion of the red planet. A view reflected in several passages in the *Iliad* is that Ares is not to be trusted, and Butler's translation says, that he is 'a villain incarnate, first on one side and then on another' (5.830), before describing him as 'Sir Facing-Bothways' (5.889). A. T. Murray uses the word 'renegade', while 'turncoat' is favoured by E. V. Rieu. Alexander Pope's 1715 translation sees Ares as defying the natural order, and declaims:

> Of lawless force shall lawless Mars complain?
> Of all the gods who tread the spangled skies
> Thou most unjust, most odious in our eyes! (5.889)

A literal translation of the word ἀλλοπρόσαλλε, which is at the heart of these variants, is 'changing from one side to another'[7]

– a truly Homeric and accurate description of retrograde motion.

In a second observation of the apparent orbit of Mars around the Earth, Homer says that Ares is 'first in front and then behind Hector' (5.595). This relatively common event takes place over a period of a few weeks, and at the beginning of the sequence Mars rises *before* Orion and *precedes* the constellation across the sky – Ares is 'first in front'. But, as the days and weeks pass, Mars rises progressively later until it rises *after* Orion and *follows* the constellation across the sky – Ares is 'then behind Hector'.

In the course of the *Iliad*, Homer's epithets associated with Ares include 'blood-stained' (5.31) and 'bloody' (5.844), both matching the appearance of the red planet as well as fitting Ares' aggressive character as a 'spear-carrying' (5.356), 'murderous' (13.802) warrior who is the 'bane of men and stormer of cities' (5.455) and looked upon by Zeus as a 'hateful troublemaker' (5.889). Today the icon or symbol used to represent Mars is still the spear. Ares was particularly disliked by Athene, who won a 'power' struggle with him at a time when Jupiter appeared brighter in the sky than Mars (5.30).

Conjunctions of two or more heavenly bodies are common events, and a number of these involving Mars are recorded in Ares' friendship with Aphrodite. After Ares has been wounded in the neck (21.415), Aphrodite, who in mythology was his wife, leads him from the scene of battle (fig. 51). A conjunction of Mars with the Moon is recorded when Ares squabbles with Hera (the Moon) (5.764), and another conjunction between Jupiter (Athene) and Mars (Ares) is noted at 15.123. There is also a conjunction of the Moon, Venus, Mars and Jupiter at 21.421.

Ares' sister, Strife, 'whose fury never tires . . . who from being first low in stature, grows till she uprears her head to heaven' (4.440), is thought to record a notable aspect of Mars during its orbit around the Sun, when it changes in brightness

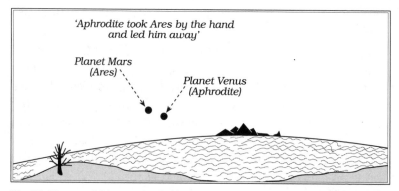

Fig. 51 *Venus and Mars setting in the west, as Aphrodite leads Ares from the battle-field.*

as its distance and position relative to the Earth vary. When Mars is moving towards maximum brightness, we propose that the planet is represented by Strife, who in Book 4 is first involved in a minor skirmish that erupts into full-scale war.

Homer uses Ares to restate his almost obsessional concern with the apparent arc of the celestial sphere (see Chapter 7), and a sequence in Book 5 shows Mars ascending to the dome of the heavens. The sequence begins when Mars rises in the east and Homer says, 'She [Iris] found fierce Ares waiting on the left of the battle' (5.355). As Mars rises in the skies, Homer says, 'Ares is now lording it in the field' (5.824), and the planet reaches its highest point at the meridian when 'As a dark cloud in the sky when it comes on to blow after heat, even so did Diomedes see Ares ascend into the broad heavens' (5.866). At this point Ares is wounded and flees to his father Zeus as the planet sets.

Mars' path through the zodiac is very close to the ecliptic, and Ares' 'monstrous spear' (5.594), like those of Athene and Agamemnon, is a symbolic representation of the path of the Sun. To draw attention to the ecliptic, Homer uses two curious literary devices. The first involves incidents when a spear thrown at a warrior misses its target and lies 'quivering' in the ground. When Aeneas, leader of the Dardanians, attacked Meriones, second in command of the Cretans, 'The spear flew past him and the point stuck in the ground, while the butt-end went on quivering till Ares robbed it of its force' (16.613). And later 'Hector . . . aimed a spear at Automedon but he saw it coming and stooped forward to avoid it, so that it flew past him and the point stuck in the ground, while the butt-end went on quivering, till Ares robbed it of its force' (17.529). For Ares twice to stop spears from 'quivering' suggested a point worthy of further investigation. In the first instance, we found that the spear thrown by Aeneas (α Virginis) at Meriones (θ Tauri) would have carried on in a direct line of sight close to a point on the ecliptic where Ares (Mars) could be seen. In the second example, the spear thrown at Automedon (β Canis Minoris) by Hector (β Orionis) would also have travelled on in a straight line towards the ecliptic and the waiting Ares.

The second curiosity is that warriors previously identified with stars on or very close to the ecliptic are killed in the same manner, by a blow to the 'belly'. In terms of magnitude these stars are not particularly bright, but they are significant in astronomical terms because of where they lie. When warriors are struck in the belly, the most central part of the anatomy, a useful mental image directs the eye to the ecliptic, the apparent path of the Sun. This helps to focus attention on Mars' closeness to the ecliptic when Ares too is wounded in the stomach when 'Pallas Athene drove the spear into the pit of Ares's stomach where his under-girdle went round him' (5.855). The incident doubly reminds the audience of the ecliptic, because Athene as Jupiter closely follows the path of the Sun.

An intriguing incident in Ares' life that has not yet succumbed to astronomical interpretation is when he was bound 'in cruel bonds, so that he lay thirteen months imprisoned in a vessel of bronze' (5.385). The text may be implying that for a period of just over a year Mars could not be seen, because it was below the horizon at night and in the sky during daylight hours.

Athene as Jupiter

Athene entered Greek culture during the Mycenaean era, *c.* 1600–1100 BC, and Homer recalls her ancient origins when he associates her with the legends of Hercules, Tydeus, Bellerophon, Erechtheus and Nestor. The daughter of Zeus, Athene was born when her father complained of a terrible headache and Hephaestus cleaved open his skull to relieve the pain. From the gaping wound stepped the adult Athene, in full armour and uttering a terrible war cry. She was rejected by the Trojan Paris when he was asked to choose the most beautiful of three female gods, but her chance for revenge came at the Siege of Troy, when, like Hera, she devoted her energies to the city's overthrow. Athene's powerful personality in the *Iliad* is matched with her astronomical role as the majestic and brilliant planet Jupiter.

It is not difficult to establish an affinity between Athene, the goddess of wisdom and war, and Jupiter, the planet that shines steadfastly at a magnitude between -1.9 and -2.4 and has a constancy that Mars lacks. Like Ares, Athene is identified with war, but, whereas he is bold and rash, she is the strategist and organizer. Jupiter moves through the skies slowly, taking 399 days to apparently orbit the Earth, and remaining in each zodiacal constellation for more than a month. Its synodic period – the time apparently taken to go round the Earth – includes 121 days when the planet appears to stop and then move into retrograde motion. Two such stationary points are recorded when Homer says Athene stands by the side of Odysseus (2.169) or puts 'valour into the heart' of Diomedes (5.1), or when the planet appears to have halted below the constellations of Boötes (Odysseus) and Perseus (Diomedes). The *Iliad* also records Jupiter in conjunction with the Moon,

Mercury, Venus, Mars and Saturn, when Athene is said to be close to Hera (4.20), Apollo (7.58), Aphrodite (21.424), Ares (5.29) and Poseidon (21.284).

An impressive sight in the skies occurs at certain times of the year when the Moon and Jupiter rise together at sunset and journey in company across the sky before setting. Athene has a close relationship with Hera, and together they hatch their plots to speed the doom of the Trojans in 2.156, 4.9 and 5.418. When the two goddesses go off together for a chariot ride, they are brought to heel after Zeus warns that if they do not return he will smash their chariot and 'it will take them all ten years to heal the wounds my lightning shall inflict upon them' (8.404). It is not difficult to equate Athene and Hera in a chariot with a conjunction of the Moon and Jupiter, but we had difficulty in interpreting the period of ten years until we studied the planet's orbit. This showed that, from a time when the full Moon and Jupiter at its brightest were both together in a particular zodiacal constellation, it will take some ten years before they meet again in the same constellation.

The 'aegis' of Athene, both in the stories of Homer and in later legend, has given rise to debate about both its purpose and its identity. It has been described as a round shield, but as Athene's legend developed it became a shield with a gorgon's head and fringed with snakes, and later a goatskin mantle embroidered with a gorgon's head and fringed with snakes. In astronomical terms, we believe the immortal aegis to be linked with the ecliptic and the orbit of Jupiter, which hugs the path of the Sun. The planet itself might be thought to give the impression of a glittering shield.

Athene, like Ares and Agamemnon, is armed with a huge spear – another concept associated with the ecliptic. Book 8 describes how she 'stepped into her flaming chariot, and grasped the spear so stout and sturdy and strong with which she quells the ranks of heroes who have displeased her' (8.389); little imagination is needed to see majestic Jupiter 'grasping' the ecliptic on its journey across the heavens. Later in Book 8, Iris, the messenger of Zeus, says of Athene, 'but you, bold hussy, will you really dare to raise your huge spear in defiance of Zeus?' (8.421). Iris' rhetorical question may imply the futility of defying Zeus with an astronomical concept, the ecliptic, under his command.

The path of Jupiter never strays far from the ecliptic, and Homer faced a challenge when he had to express Athene's affection for or assistance to the warriors Achilles, Odysseus and Diomedes, whose personal stars lie at a good distance *outside* the zodiac. In the case of Achilles (α Canis Majoris), Homer says that Athene supports him 'from afar' (18.217), or that 'of the others no man could see her' (1.197). When she 'moves up close' to Odysseus (α Boötis), Jupiter lies in the zodiac below his feet (2.169). In another instance (5.120) Athene needs to move into the zodiac south of Perseus, whose feet only just reach into the zodiac, so she can hear Diomedes' prayers.

However, Athene, has no difficulty when involved in the affairs of warriors whose stars and constellations lie *in* the zodiac. As Jupiter travels through the broad-shouldered and long-legged constellation of Scorpius, Athene's infusion of fighting spirit into Menelaus is described with the well-known simile of the fly: 'Therefore she put strength into his knees and shoulders and made him as bold as a fly, which, though driven off will yet come again and bite if it can' (17.569). On other occasions she assumes the voice or form of a mortal when Jupiter occults or temporarily covers a star. She takes the form of Laodocus, λ Virginis, to persuade the Trojans that they should kill Menelaus (4.86). After Patroclus' death she takes the form of Phoenix, β Cancri, to encourage Menelaus to guard Patroclus' body (17.555). An occultation, and the treachery associated with it, leads to the death of Hector when Athene assumes the form and voice of his brother Deiphobus, ζ Tauri. When Athene speaks as Deiphobus she says that 'he' and Hector will fight Achilles together (22.227). However, when

Hector calls upon his brother for support he is not there – Jupiter has obscured Deiphobus' personal star. Hector says, 'Alas! the gods have lured me on to my destruction. I deemed that the hero Deiphobus was by my side . . . Athene has inveigled me' (22.297).

Aphrodite as Venus

Aphrodite needs no introduction as the goddess of love and beauty who was later, like the planet, named Venus by the Romans. 'Laughter-loving' (4.10), she is the most feminine of all the goddesses, and a marked contrast to the belligerent Hera and Athene. The daughter of Zeus and Dione, she is the mother of the mortal Aeneas, leader of the Dardanians, whose life she helps to save in astronomical allegory (5.311).

Venus can be seen shining brilliantly for up to three hours at dusk or dawn, and to the ancients the planet was known as the 'evening star' when seen in the west and the 'morning star' when seen in the east: 'At length . . . the Morning Star was beginning to herald the light which saffron-mantled Dawn was soon to suffuse over the sea' (23.226).

When Venus is positioned in the zodiac above Orion, Aphrodite keeps a protective watch over Paris and Hector, sons of King Priam. She rescues terrified Paris from death at the hands of Menelaus: 'Aphrodite snatched [Paris] up in a moment . . . and conveyed him to his own bedchamber' (3.380). In this instance, Homer draws the evening sky as Venus (Aphrodite) and the constellation Orion (Paris) both dip below the western horizon, while Scorpius (Menelaus) rises in the east (see Chapter 7). Similarly, Venus is above Orion when Aphrodite protects Hector's body after he is slain by Achilles: 'but the dogs came not about the body of Hector, for Zeus's daughter Aphrodite kept them off him night and day, and anointed him with ambrosial oil of roses' (23.185).

A conjunction of Venus and Mars as they are setting occurs in

Book 21, when 'Queen Hera saw . . . that vixen Aphrodite . . . taking Ares through the crowd out of the battle' (21.416). The Moon and Venus were in the sky just before the lunar eclipse during the famous seduction of Zeus, when Hera tried to enlist the help of the goddess of beauty: '"My dear child," said [Hera], "will you do what I am going to ask of you, or will you refuse me because you are angry at my being on the [Greek] side, while you are on the Trojan?"' (14.190). Venus then disappeared from the evening sky and the Moon (Hera) took central stage when 'Aphrodite now went back into the house of Zeus, while Hera darted down from the summits of Olympus' (14.224).

Venus can also go into retrograde motion, and one of the occasions on which it appears to be stationary before resuming its journey is thought to be when Aphrodite came to the house, Orion, of Paris: 'the laughter-loving goddess took a seat and set it for her facing Paris' (3.424).

Apollo – Mercury at Sunrise

Mercury, the planet closest to the Sun, is visible to the naked eye at dawn or dusk for less than two hours, for only short periods during the year. As bright as a first-magnitude star, the planet never rises far above the horizon and is easily obscured by low cloud or mist, making it difficult to find. Copernicus is reputed never to have seen it from his observatory in Frauenburg in Poland.

Phoebus Apollo has long been associated with light, and as Mercury in the morning skies he heralds dawn and the rising Sun. Later mythology associates Apollo with the Sun, but in Homer the god has, for a number of reasons, been linked to the planet Mercury when it is seen before the Sun rises. In Book 10, during the night raid on the Trojan lines (10.518), Apollo wakens the Thracian Hippocoön shortly before dawn, the time when Mercury can be seen. Apollo is another supporter of the Trojans and their allies, and as Mercury he champions them as

178

the planet travels through the zodiac. With Mercury above Orion, perhaps in Taurus, Apollo is said by Homer to be aiding Hector in his clashes with Aias, Teucer, Diomedes and Achilles. When Mercury is in the zodiacal constellation of Gemini, Apollo helps Glaucus, who leads the men from Lycia.

Mercury in Virgo takes part in an unusual occultation when twice in succession Spica, α Virginis, the personal star of Aeneas, is obscured by or appears to come very close to planets. On the first occasion, Aphrodite (Venus) covers her son, Aeneas (Spica), with a veil and begins to whisk him away to safety, after he is badly wounded. She in turn comes under attack and drops the unfortunate Aeneas. Mercury, however, is close by and Spica (Aeneas) is occulted when 'Phoebus Apollo caught him [Aeneas] in his arms, and hid him in a cloud of darkness' (5.344). On a separate occasion (21.545) with Mercury in Virgo, an occultation is recorded when Apollo takes the form of Agenor (γ Virginis).

Mercury is in conjunction with other planets when narrative associates Apollo with Hera (the Moon), Aphrodite (Venus), Athene (Jupiter), Ares (Mars) and Poseidon (Saturn). The only planet-god never in the sky at the same time as Apollo is Hermes, who represents Mercury as an evening star. As the archer-god whose arrows wrought havoc on the Greek forces at the beginning of the *Iliad* (1.53), Apollo has a second role as the constellation Sagittarius.

Hermes – Mercury in the Evening

When seen in the evening sky, Mercury is represented by Hermes, the swift messenger of the gods and the guardian of travellers. It is in the latter role that he has his moment of prominence in Book 24 of the *Iliad*. Hector has been killed and Patroclus has been honoured in funeral games, but Achilles' rage has still not been assuaged and he continues to defile Hector's body. Traditionally, Hermes is also a thief and trickster

– attributes recalled by Homer when the god is called upon in vain to steal Hector's body from Achilles so that it can be returned to his father, King Priam (24.24). Then Hermes assumes his role as a guide after Priam sets out from Troy (Ursa Major) one evening to plead with Achilles (Canis Major) for the return of Hector's body. He is met on the journey by Hermes (Mercury in the evening sky), but the god can take him only part of the way. Priam's 'journey' establishes the relative positions of Ursa Major, the zodiac and Canis Major. The sequence is: Priam leaves Ursa Major, and is taken across the zodiac by Hermes (Mercury) before meeting Achilles in Canis Major. After paying a ransom for Hector's body, Priam is urged to return home by Hermes while it is still dark.

The first hurdle that Priam meets on his outward journey is the zodiac, where Hermes is instructed by Zeus, 'It is you who are the most disposed to escort men on their way, go and so conduct Priam to the ships of the Greeks' (24.334). As a planet, Hermes cannot move outside the boundaries of the zodiac, and after crossing the zodiacal band he takes his leave of Priam when he says, 'I will now leave you [Priam], and will not enter into the presence of Achilles' (24.462). Although it would not be possible to see Mercury both in the evening and in the morning in the course of one night, Homer appears to reinforce the position of the planet in the zodiac when Priam, returning home, is again met by Hermes and is escorted to the topmost boundary of the zodiac, from where Priam has to continue his journey without the god's help (24.682).

Hephaestus as Perseus

A study to discover the astronomical role of the smith-god Hephaestus led down a path with a surprising end. Every reference to the great artificer was studied, and, by using techniques similar to those used for other interpretations, an astronomical image of the constellation Perseus emerged. It was also apparent that, in addition to his role as creator of the constellations, Hephaestus had a further role of previously unsuspected importance.

'Then Hephaestus drew sweet nectar from the mixing-bowl, and served it round among the gods, going from left to right; and the blessed gods laughed out a loud applause as they saw him bustling about the heavenly mansion . . . But when the sun's glorious light had faded, they went home to bed, each in his own abode, which lame Hephaestus with his consummate skill had fashioned for them' (1.597). This quotation is the first indication of Hephaestus' role as the creator of the constellations as 'abodes' for the gods, and his serving of nectar from left to right indicates the direction in which constellations appear to move from east to west across the skies.

The constellations in the southern hemisphere were built by Hephaestus when he was flung down from Olympus to Lemnos. 'He [Zeus] caught me by the foot and flung me from the heavenly threshold. All day long from morn till eve, was I falling, till at sunset I came to ground in the island of Lemnos, and there I lay . . . It would have gone hardly with me had not Eurynome . . . and Thetis, taken me to their bosom' (1.591 and 18.395). There, in the realms of Thetis and her home in the constellation Eridanus below the horizon, Hephaestus stayed for nine years, creating the southern constellations – 'beautiful works in bronze, brooches, spiral armlets, cups, and chains . . . in their cave' (18.400).

It is as the creator of the glittering new armour and shield of Achilles that Hephaestus is perhaps best known (see Chapter 7), and his masterpiece represents the stars in an area of sky where a new constellation was created to accommodate the return to the heavens of the star Sirius after a long absence. In his role as the creator of astronomical concepts, Hephaestus is also credited with making the ancient staff that had been passed down to Agamemnon by his forebears – a symbolic pointer to the north celestial pole.

To provide such astronomical functions, Hephaestus required his own home in the skies, and the details given in Book 18 when assessed in the same way as the information about other gods and warriors resulted in a vivid and unmistakable image of Perseus. How Hephaestus could represent Perseus if the constellation had already been allocated in one manifestation to Diomedes (see Chapter 5) is not at present understood, but an answer may yet be found in pages of notes written by Edna Leigh. In these she associates many other Greek gods with constellations, and we suspect these were her conclusions on yet another layer of ancient Greek astronomy. No explanatory notes for her views have yet been found, but it is expected that her papers will be made available for those who wish to study the deeper implications of *Homer's Secret Iliad*. Considering Hephaestus' role as the original creator of the constellations, new lines of inquiry may indicate that Homer preserved in the *Iliad* the remnants of an even older astronomical culture than that associated with Minoan Crete and Mycenae.

Consider the following two passages:

Meanwhile Thetis came to the house of Hephaestus, imperishable, star-bespangled, fairest of the abodes in Heaven, a house of bronze wrought by the lame god's own hands. She found him busy with his bellows, sweating and hard at work, for he was making twenty tripods that were to stand by the wall of his house, and he set wheels of gold under them all that they might go of their own selves to the assemblies of the gods, and come back again – marvels indeed to see. (18.368).

The mighty monster hobbled off from his anvil, his thin legs plying lustily under him. He set the bellows away from the fire, and gathered his tools into a silver chest. Then he took a sponge and washed his face and hands, his shaggy chest and brawny neck; he donned his shirt, grasped his strong staff, and limped towards the door. There were golden handmaids also who worked for him, and were like real young women, with sense and

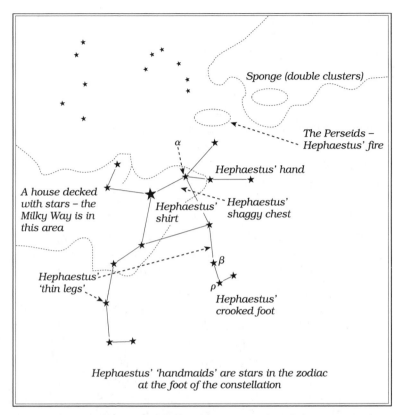

Fig. 52 The constellation Perseus as Hephaestus. Hephaestus' twenty tripods are the twenty bright stars of the constellation.

reason, voice also and strength, and all the learning of the immortals. (18.410)

The opening words of the first passage place the home of Hephaestus firmly in the sky, and the twenty tripods he is making match the twenty or so bright stars in the outline of the constellation Perseus. The golden 'wheels' may represent stars as they move across the skies.

In Book 1 the gods laugh at lame Hephaestus (1.599), and in the second quotation his deformity is shown again when he hobbles from his anvil on his thin legs. A glance at fig. 52

shows that, although Perseus has two long legs, one of them is shorter than the other – a positive reminder of the lameness of Hephaestus. When the smith-god hobbles, Homer appears to be drawing attention to two variable stars on the foot of the short leg of the constellation. On the 'ankle' is the notable Algol, β Persei, also known as the Demon Star. For most of the time Algol has magnitude 2.2 but over a few hours almost every three days it fades to magnitude 3.4, before regaining its usual brightness. On the 'heel' of Perseus is a second naked-eye variable star, ρ Persei, which varies between magnitudes 3 and 4 during a period of between thirty-three and fifty-five days. Hephaestus' strong staff represents the Milky Way usually depicted as Perseus' sword, while his 'shaggy chest' and 'tools' in a 'silver chest' are stars in the middle and lower part of the constellation that stand out against the Milky Way. The sponge with which he washes his face consists of two double clusters, each about the size of the full Moon, in the higher part of Perseus, while his shirt is a common Homeric word to identify four bright stars that make a rectangular asterism in the centre of the constellation. The 'golden hand-maidens' with their suggestion of immortality are stars in the zodiac just below the lower foot of Perseus.

When the mighty Xanthus, or the Milky Way, is sickened by Achilles' wanton killings, he rises up against him and Achilles is in great danger of drowning. Hera, knowing that Hephaestus has a 'fire', calls out, 'Crook-foot, my child, be up and doing, for I deem it is with you that Xanthus is fain to fight; help us at once, kindle a fierce fire . . . that shall bear the flames against the heads and armour of the Trojans and consume them, while you go along the banks of Xanthus burning his trees and wrapping him round with fire' (21.331). Astronomically, this suggests the Perseids meteor shower with its radiant in the Milky Way, and Homer is presenting an image of the Milky Way being set on fire by the Perseids (see Chapter 5, page 146).

Thetis and her Nymphs: Eridanus

Silver-footed Thetis and the Nereids, or sea nymphs, are the constellation Eridanus (the River), most of which lies in the 'depths of the sea' (below the horizon) from the latitude of Greece. Thetis' personal star is Achernar, α Eridani, a bright star positioned at the foot of the constellation, far in the southern hemisphere, and which gives the goddess her epithet of 'silver-footed'.

Thetis is the mother of Achilles, and when she rises from the waves in Book 18 the long list of her attendant nymphs seems meaningless and even tedious until linked to the stars of the meandering constellation:

> Then Achilles gave a loud cry and his mother heard him as she was sitting in the depths of the sea . . . whereon she screamed, and all the goddesses daughters of Nereus that dwelt at the bottom of the sea, came gathering round her. There were Glauce, Thalia and Cymodoce, Nesaia, Speo, Thoe and dark-eyed Halie, Cymothoe, Actaea and Limnorea, Melite, Iaera, Amphithoe and Agave, Doto and Proto, Pherusa and Dynamene, Dexamene, Amphinome and Callianeira, Doris, Panope, and the famous sea-nymph Galatea, Nemertes, Apseudes and Callianassa. There were also Clymene, Ianeira and Ianassa, Maera, Oreithuia and Amatheia of the lovely locks, with other Nereids who dwell in the depths of the sea. (18.34)

Edna Leigh wrote:

> I believe we can identify [the nymphs] as individual stars of Eridanus presented in straightforward sequence, beginning with the one we call α and ending with ω . . . Homer appears to be using the stars nearest Orion in another passage. [The star in more modern times designated as β Eridani but lying close to Orion has been defined in Homeric astronomy as Cebriones, Hector's charioteer. See Chapter 4, page 105, and fig. 23.]
>
> For some of the names I am suggesting a short translation. It will be readily noted from the list of nymphs that the word *amphi* which suggests 'two' – literally 'on both sides' – occurs twice: once with

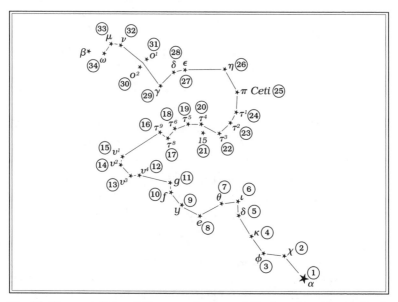

Fig. 53 *The constellation Eridanus as the home of Thetis and her Nereids 'in the depths of the sea'. This diagram is derived from one sketched by Edna Leigh.*

Amphithoe and once with Amphinome. Each time it will be seen that two stars are indicated, or perhaps a remarkable coincidence allows only an apparent correlation. Six nymphs are given twin names: Doto and Proto; Dynamene and Dexaneme, both, as it happens, of magnitude 4.3; and Ianeira and Ianassa, listed in modern star catalogues as o^1 and o^2, two stars close together that can be seen by the naked eye.

Including Thetis, Homer tells of thirty-four nymphs. For the sake of clarity, these are numbered in Table 10 and can be identified on the drawing of Eridanus in fig. 53. The star list begins in the far south with Thetis (Achernar, α Eridani), and 'flows' northward to ω Eridani near the right knee of Orion.

Edna Leigh added:

Eridanus is an old constellation named as a rule by ancient peoples after a local river. It was impossible for Homer to know of

Table 10 *Thetis and her nymphs in Eridanus*
(see fig. 53)

1.	α	Thetis	18.	τ^6	Pherusa
2.	χ	Glauce	19.	τ^5	Dynamene
3.	φ	Thalia	20.	τ^4	Dexamene
4.	κ	Cymodoce	21.	15	Amphinome
5.	σ	Nesaia	22.	τ^3	Callianeira
6.	ι	Speo	23.	τ^2	Doris
7.	θ	Thoe	24.	τ^1	Panope
8.	e	Halie	25.	π	(Cetus) Galatea
9.	y	Cymothoe	26.	η	Nemertes
10.	f	Actaea	27.	ε	Apseudes
11.	g	Limnorea	28.	δ	Callianassa
12.	v^4	Melite	29.	γ	Clymene
13.	v^3	Iaera	30.	o^2	Ianeira
14.	v^2	Amphithoe	31.	o^1	Ianassa
15.	v^1	Agave	32.	ν	Maera
16.	τ^9	Doto	33.	μ	Oreithuia
17.	τ^8	Proto	34.	ω	Amatheia

Achernar by direct observation from either Greece or Asia Minor, for during his time the star lay near the South Pole. Nevertheless, Homer does know the constellation, but how he acquired such knowledge will be a matter of some interest.

Signs, Seers and Astrology

As we noted in Chapter 1, when the observers of ancient Babylon recorded natural events they had seen in the heavens, they were not concerned with the pursuit of astronomy for its own sake but were seeking signs or omens. Their belief in astrology has been shared by societies throughout the world, in both the ancient past and the present. It is therefore not unreasonable to conjecture that the Greeks of Homer's times might also have sought to predict the future from the skies.

Certainly Homer was aware of the arts of soothsaying, for he refers to omens and signs on a number of occasions, and the cast of the *Iliad* includes Calchas, the 'seer and diviner of omens' (1.68), and Ennomus 'the augur' (2.858), while

Helenus, brother of Hector, is the 'wisest of augurs' (6.75). Omens interpreted from the behaviour of birds are referred to on a number of occasions, and the phrases 'birds of ill omen' and 'birds of omen' are used. On one occasion Homer describes in dramatic detail an episode involving an eagle and a snake, but, whatever dread foreboding it might have brought into the minds of ancient Greeks, this also served the astronomical purpose of switching attention from one part of the sky to another. This event occurs amid furious fighting in Book 12, when a monstrous eagle soars across the battlefield and drops a blood-red snake among the Trojan ranks (12.200). The bird represents the Greek constellation of Aquila (the Eagle), and the snake represents Serpens Cauda ('Trojan' stars in the Serpent's Tail), around which skirmishes had been taking place. To change the scene from one end of the skies to another, Homer says the eagle, with the snake in its talons, flies skirting the left (eastern) wing of the Trojan host before dropping its victim into the ranks of Trojan warriors from Orion and Gemini in the west.

Perhaps the nearest Homer comes to describing an astronomical event as an omen is in Book 4, when:

> Athene darted from the topmost summits of Olympus. She shot through the sky as some brilliant meteor which the son of scheming Cronus has sent as a sign to mariners or to some great army, and a fiery train of light follows in its wake. The Trojans and Achaeans were struck with awe as they beheld, and one would turn to his neighbour, saying, 'Either we shall again have war and din of combat, or Zeus the lord of battle will now make peace between us.' (4.73)

It is significant that Homer does not say Athene's meteor-like appearance *was* an omen, only that it was similar to one. He is careful not to specify exactly what such an omen would mean, and for warriors to say that such a sign could have meant war *or* peace is positively Delphic in its ambiguity. Lightning and rainbows are also referred to as omens, but, as with other references to Zeus sending signs to mortals, there is a lack of detail about their significance.

Homer was well aware of the soothsaying and divination beliefs of his era, and even the most general references to birds of ill omen and other signs would have struck a familiar chord with his audiences. Specific astrological prognostications are notable in the *Iliad* for their apparent absence, and when Homer writes about the heavens in allegory he stays true to his purpose of preserving only astronomical knowledge.

7

The Changing Heavens and the Fall of Troy

> King Priam was first to note [Achilles] as he
> scoured the plain, all radiant as the star
> which men call Orion's Hound, and whose
> beams blaze forth in time of harvest more
> brilliantly than those of any other that
> shines by night.
>
> *Iliad* 22.25

The devising of constellations, one of the first astronomical
achievements in prehistory, must have gone hand in hand
with observations of the movement of the heavens. The minds
of ancient peoples were inquiring and reasoning, and they
sought explanations for the phenomena they observed.

At some unknown time and place, prehistoric watchers of
the skies began to put their observations into a logical frame-
work, to bring reason to the movement of the heavens. This
was the dawn of theoretical science, and it took a tremendous
leap of intellect to create abstract ideas from what could be
seen in the ever-changing skies over weeks, months and many
a long year. The nightly sight of constellations rising in the east
and ascending in an arc before setting in the west raised questions about the nature of the universe. At some time in history
it might have been noticed that a new star had appeared on the

horizon, or that the star that indicated celestial north was not as accurate as it once had been. These and other observations led ancient people to formulate two of the fundamental ideas of astronomy: the precession of the equinoxes and the place of the Earth in the universe. The first of these is discussed in this chapter, and the second in Chapter 8.

To recapitulate briefly on what has already been set out in Chapter 3, Homer recorded three observable effects of precession. In the order in which they are presented in this chapter, they are:

1. Stars that for ages of time could not be seen at a particular latitude come into view on the horizon. The return to the skies of Greece of Sirius, α Canis Majoris, is described in the allegory of the return to the battlefield of Achilles, the *Iliad*'s most powerful warrior.
2. The constellations in which the Sun rises at the vernal and autumnal equinoxes change over long periods of time. Homer records the passage of the autumnal equinox through three zodiacal constellations, and similarly the vernal equinox.
3. Changing pole stars. The point where an extension of the Earth's axis would meet the celestial sphere is called the celestial pole, and the star nearest to it at any one time is called the pole star. This point is not fixed, however, and over long periods of time different stars become the pole star. The decline and fall of Troy is an allegory associated with the decline of Thuban, α Draconis, as the north pole star from *c.* 2800 BC to *c.* 1800 BC.

Precession is said to have been discovered by the Greek scientist Hipparchus (fl. 135 BC), who lived six hundred years after Homer, but another 1,800 years elapsed before Sir Isaac Newton explained the phenomenon in modern scientific terms. In writing about Hipparchus, Pliny the Elder (AD 23–79) recorded the story that, having seen an unspecified new star in the sky, the Greek scientist questioned 'whether the stars that we think to be fixed are also in motion'.[1] It might be asked if indeed a new star did appear in Hipparchus' time, as Pliny recorded, or whether he was advancing new ideas on an observation made in the distant past.

The *Iliad* shows that, although Homer might not have known the scientific reasons for precession, he was certainly aware of the important changes it brought about.

The Return of Sirius

At a time far away in history, *c.* 8,900 BC, it would have been observed from the latitude of Greece that a brilliant new star – the brightest in the sky – had appeared low on the horizon; the shifting heavens had brought Sirius, α Canis Majoris, back into view. How Homer, who lived more than eight thousand years later, knew of this event is not known, but so graphic is his exposition of the return of Sirius in the allegory of Achilles that he undoubtedly did know of it.

The *Iliad* opens with a furious argument over a slave-girl between Achilles and his commander, King Agamemnon. Achilles has to give up the girl to his leader, but in retaliation he says he will no longer fight for the Greeks and goes away to sulk. As he leaves he warns, 'Solemnly do I swear that hereafter they shall look fondly for Achilles and shall not find him' (1.240). Many pages of fighting later, the war is not going well for the Greeks and Agamemnon pleads with Achilles for help in stemming the Trojan onslaught. Achilles rejects Agamemnon's approach, but when his friend Patroclus is killed by Hector he returns to seek his revenge. Even then his reappearance is delayed while the smith-god Hephaestus forges him a glittering new suit of armour and a shield, and by negotiations with Agamemnon over the gifts Achilles is to receive. With the Greek forces ready to herald the return of their hero, there is yet further delay when Odysseus suggests the army should pause and restore themselves with a meal before fighting resumes.

When those literary events are seen as allegories, a remarkable astronomical cycle is revealed. Achilles' departure from the battle is a reminder that for thousands of years Sirius, his personal star, could not be seen from Greece. Achilles could not have been more precise when he left the 'battlefield' and said that henceforth the Greeks would not find him with them: Sirius did not rise above the horizon in Greece for some seven thousand years before its reappearance *c*. 8900 BC.

Homer builds the tension when he takes almost two books or chapters to prepare Achilles for his return to the literary battlefield – a reflection of the millennia that passed before Sirius returned to the sky. His mother, Thetis, commissions from Hephaestus, the creator of the constellations, an amazing new shield and a suit of glittering armour:

> First [Achilles] put on the goodly greaves fitted with ankle-clasps, and next he did on the breastplate about his chest. He slung the silver-studded sword of bronze about his shoulders, and then took up the shield so great and strong that shone afar with a splendour as of the moon . . . He lifted the redoubtable helmet, and set it upon his head, from whence it shone like a star, and the golden plumes which Hephaestus had set thick about the ridge of the helmet, waved all around it. (19.369)

Astronomically, the armour matches the outstanding beauty of Sirius and stars in Canis Major, while the shield represents the region of the sky in which the new star was placed.

It has been suggested it would have been more in the heroic mould if Achilles, decked in his shining new armour, had rushed immediately into battle while full of sorrow and anger at the death of Patroclus. To stress that there had been a very long wait in astronomical terms for Achilles/Sirius to rejoin the battle of Troy in the heavens, Homer delays the fighting still further while the Greeks debate whether or not the troops should first take breakfast.

Homer confirms in Book 19 that Achilles is an astronomical newcomer when he is compared to Odysseus, whose constellation, Boötes, had been in the skies for an eternity. Odysseus says, 'Achilles . . . mightiest of all the Greeks in battle, you are better than I, and that more than a little, but in counsel I am much before you, for I am older and of greater knowledge' (19.216). Homer is here stating that Odysseus' personal star, Arcturus, α Boötis (magnitude −0.04), is not as bright and powerful as that of Achilles, Sirius, α Canis Majoris (magnitude −1.4), but is much 'older', for it has been visible from the northern hemisphere since time immemorial.

The return of Sirius to the northern skies did not happen suddenly, overnight, and no precise year can be given for when the star was first seen again in Greece. The further south the latitude of the observer, the earlier Sirius would have appeared. People in Crete would have seen it return to the night sky before those living further north in Athens, and it would also have made a difference if the observer were standing on a hilltop or at sea level. The date quoted in this work of c. 8900 BC is an approximation based on a computer projection for the latitude of 38° north – within a degree of the latitude of Athens and of Homer's possible home in Chios or Smyrna (Izmir).

The appearance of the brightest star was not only a major spectacle: it also raised the problem of how to incorporate Sirius in the then established order of constellations. In the *Iliad* this was solved by a major revision of stars and constellations in a large sector of the sky.

In an area of the heavens extending south towards the horizon and bounded in the north by Cancer, in the west by the Milky Way and Orion and in the east by Hydra there were only a small number of significantly bright stars before the arrival of Sirius (fig. 54). These included the stars now known as

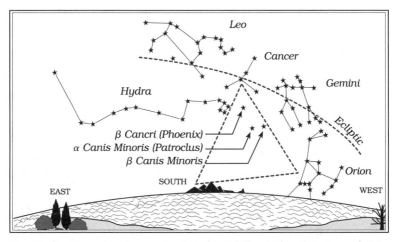

Fig. 54 *Creating a kingdom in the skies for Achilles. Before the return of Sirius c. 8900 BC, the only particularly bright star in the dashed triangle that would have met the criteria for Phoenix would have been β Cancri.*

Procyon, α Canis Minoris, and Altarf, β Cancri, which may have been part of an extended ancient constellation of Leo. A reorganization of this area of sky to accommodate Sirius is indicated by the importance Homer assigns to the 'gifts' offered to Achilles if he will return to the battlefield. On three different occasions the gifts are listed in detail as seven tripods, twenty iron cauldrons, twelve horses and eight women. If each group of gifts is taken to represent stars within a band of similar brightness or magnitude, it suggests that some stars within an older boundary of Leo were 'given' to a new constellation boundary created around Sirius and including α Canis Minoris and β Cancri, perhaps once 'owned' by Agamemnon (Leo). Some support for the idea that a more ancient configuration of Leo was larger than the present constellation is derived from a belief that the earliest zodiac had only six divisions and not the twelve devised later.[2]

It is significant that Achilles and Patroclus are also said to be sitting opposite each other in the same 'hut' or 'tent', in just the

195

same way that their personal stars, Sirius and Procyon, would have sat 'opposite' each other within the same constellation boundary. In a literary sense this has been thought to imply a particularly close relationship between the two companions.

Another incident in the *Iliad* where 'gifts' are thought to represent stars occurs when King Priam of Troy pleads with Achilles for the return of Hector's body. As ransom for the corpse he takes twelve robes, twelve cloaks, twelve rugs, twelve mantles, twelve shirts, ten talents of gold, two burnished tripods, four cauldrons and a beautiful cup (24.228). Edna Leigh believed that Homer was cataloguing the stars of Ursa Major, and assigned the robes, cloaks, rugs, mantles and shirts to sixty of the constellation's fainter stars. For the remaining seventeen objects of greater value, she proposed that the tripods were the brightest stars in the constellation, with magnitudes of 1.9 to 1.7; the cauldrons were the four stars of the bowl of the Big Dipper, with magnitudes of 2.0, 2.4, 2.5 and 3.4; and the talents were stars ranging in brilliance from 3.1 to 3.9. 'Cups' in Homeric astronomy generally represent two stars very close together, and Priam's cup was designated as the pair of stars Alcor and Mizar in the handle of the Dipper.

When Agamemnon's gifts – or stars – were received by Achilles, Homer's allegory of the creation of a new constellation was complete. A reallocation of stars in this manner is suggested in Book 9, in the story which Achilles' friend Phoenix tells when he, Great Aias and Odysseus visit Achilles to try to persuade him to return to the fighting (9.168). The role of Phoenix has caused many problems for classicists, and in his commentary on the *Iliad* Malcolm M. Willcock devotes the best part of two pages to examining various possibilities.[3] The difficulty appears to lie in translation, and whether or not two commanders (Odysseus and Great Aias) or three commanders (Odysseus, Great Aias and Phoenix) journey to the hut of Achilles. After Great Aias and Odysseus return to Agamemnon's camp, Phoenix stays behind and is later named as a captain of Achilles' troops. In astronomical terms there is a logical answer to the mystery of Phoenix, and it appears that he did not need to 'travel' to the hut – or constellation – of Achilles, because he was already there – or very close – as the star β Cancri.

Phoenix tells a rambling story of his ancestry which suggests an earlier astronomical order when he was included in a different grouping of stars. The wrangling of Phoenix's parents and his departure from home suggest a time when he had to leave his 'family' constellation for pastures new. Little is known about Phoenix except (a) he tells of leaving his father's home; (b) he went to live in the house of Achilles' father, and was a mentor to the young boy; (c) his form is later assumed by a god; (d) he is named as a captain in the regiment of Achilles. An astronomical interpretation of this information indicates that (a) while Phoenix's star would remain in the same place in the sky, a redrawing of boundaries could allocate it to a different constellation; (b) Phoenix's personal star is in the same area as Canis Major and Canis Minor; (c) to have his form assumed by a god, Phoenix must be a star which is occulted by or seems to come very close to a planet, and is therefore in the zodiac; (d) as a captain of the Myrmidons, his star would be within the boundary of the new constellation created to accommodate Achilles. The only notable star that meets these criteria in that part of the sky is β Cancri, which is only just within the modern constellation boundaries of Cancer and close to Canis Minor.

Redrawing the heavens has not been confined to prehistoric times but has continued into the twentieth century. A striking example of the reorganization of an ancient constellation concerns Argo Navis, the great ship, which was once the largest single grouping of stars in the heavens. In AD 1750 it was redrawn by the French astronomer Louis de Lacaille and divided into three more manageable parts: Vela (the sails), Puppis (the stern) and Carina (the keel).

With the new constellation that included Sirius created, Homer had to define its boundaries and place it in the heavens. To do this he turned to Hephaestus and the allegory of the great shield of Achilles.

Achilles' Shield and Canis Major

The creation of Achilles' shield is one of the most beautiful pieces of narrative in the *Iliad*, and, as Homer must have intended, it is not easily forgotten. The importance of the shield – which bears little resemblance to any shield ever used by earthly warriors – can be gauged from the 125 lines or so of narrative used to describe its features, but why the shield looked as it did is one of the mysteries of the epic. As with so much else in the *Iliad*, however, an answer to this long-standing mystery can be found in the heavens, and the need to create a new constellation in a previously undistinguished part of the sky to accommodate the return of brilliant Sirius. There could be no better way to describe that area of sky than by extended metaphor, using images not encountered elsewhere in the epic.

Homer begins by associating the shape of the shield with the celestial sphere, upon which he places the Sun, the Moon, the Milky Way and a handful of stars – but not Canis Major, the eventual home of Achilles. He then goes on to describe wonderful decorations on the shield, including two warring cities, judges and disputing litigants, a wedding scene, reapers and binders, a vineyard and fallow land, cattle and lions, herdsmen, dogs and dancers. Not only has Homer identified the region of the sky in which the new constellation is to be created (see fig. 55), but each scene of rural or town life is believed to describe separate parts of the sky which, when seen as a whole, encompass the constellations now known as Canis Major, Canis Minor and Monoceros (see fig. 54). Homer could not include these constellations in his opening remarks on the

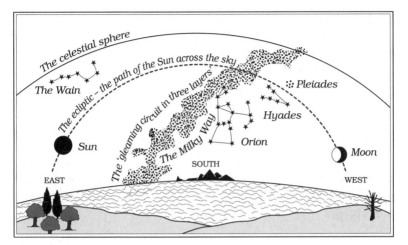

Fig. 55 *In describing how Hephaestus created the great shield of Achilles, Homer referred to the Sun, the Moon, the Milky Way and the stars shown in this diagram.*

shield (see below), for the shield had not yet been finished and the stars had not been described.

HEPHAESTUS' VIEW OF THE HEAVENS

Homer's description of the shield begins:

> First [Hephaestus] shaped the shield so great and strong, adorning it all over and binding it round with a gleaming circuit in three layers; and the baldric was made of silver. He made the shield in five thicknesses, and with many a wonder did his cunning hand enrich it. He wrought the Earth, the heavens, and the sea; the Moon also at her full and the untiring Sun, with all the signs that glorify the face of Heaven – the Pleiads, the Hyads, huge Orion, and the Bear, which men also call the Wain and which turns round ever in one place, facing Orion, and alone never dips into the stream of Oceanus. (18.487)

That these words refer to the skies is not disputed by scholars, but no definitive reason has ever been given for the more unusual phrases. Our proposed interpretation is that the 'gleaming circuit in three layers; and the baldric . . . made of silver' are the Milky Way, whose three layers are the bright central section flanked on both sides by fainter bands of light. The silver baldric is a metaphor for the way in which the Milky Way encircles the Earth, just as a strap fastened to a shield goes round the body of a warrior. Homer sets the scene in the northern hemisphere by naming the Wain (or Plough) in Ursa Major, and by identifying Orion, the Hyades and the Pleiades he uses three well-known groups of stars to determine the region of sky into which Sirius is to be placed. The Milky Way runs from top to bottom close to Canis Major, but the significance of the five 'thicknesses' is not yet clear.

DECORATION OF ACHILLES' SHIELD AS CANIS MAJOR

Homer then changes the mood, introducing scenes quite different from anything else in the *Iliad*, using images of rural and town life as metaphors for individual stars, groups of brighter stars, dark parts of the sky with few stars, and also the Milky Way. As Hephaestus carefully constructs the shield, Homer is describing in detail, and section by section, the skies that became the home of newly risen Sirius. That Homer considered Canis Major and its environs to be important is reinforced by the different ways he describes those stars in other parts of the *Iliad* (see the section on Achilles in Chapter 5).

A difficulty in proposing this interpretation is that the precise boundaries of the constellations being described by Homer are not known. He does give some help, inasmuch as 'codewords' are used in this passage in the same way in which they are used elsewhere in the *Iliad* – particularly when the Milky Way is being described in rural terms. These codewords can evoke the multitude of pinpoints of light in the Milky Way, and include the 'wheatfields' and 'vineyards' in which 'swathe after swathe' of 'harvest corn' are cut and 'grapes' are grown 'on silver poles'. The Milky Way is also described as a 'porridge of white barley'. Groups of people such as heralds, dancing youths, reapers and binders indicate stars that stand out against the background of the Milky Way. More specific individuals such as judges, look-outs, women at their doorways, the owner of the land, four cattle, nine dogs, a bull and two lions are suggested as brighter individual stars. Homer also mentions colours such as gold and silver, and these could be linked to yellow and white stars.

THE TWO CITIES

He wrought also two cities, fair to see and busy with the hum of men. In the one were weddings and wedding-feasts, and they were going about the city with brides whom they were escorting by torchlight from their chambers. Loud rose the cry of Hymen, and the youths danced to the music of flute and lyre, while the women stood each at her house door to see them.

Meanwhile the people were gathered in assembly, for there was a quarrel, and two men were wrangling about the blood-money for a man who had been killed, the one saying before the people that he had paid damages in full, and the other that he had not been paid. Each was trying to make his own case good, and the people took sides, each man backing the side that he had taken; but the heralds kept them back, and the elders sat on their seats of stone in a solemn circle, holding the staves which the heralds had put into their hands. Then they rose and each in his turn gave judgement, and there were two talents laid down, to be given to him whose judgement should be deemed the fairest. (18.490)

This first city is Sirius, and Homer describes stars in the northern part of Canis Major (fig. 56). 'The hum of men', 'weddings and wedding feasts', brides 'escorted by torchlight' and youths dancing conjure up a busy image of the Milky Way to the east. Elders sitting in a circle of polished stones represent the circular pattern of stars surrounding Sirius, and the two litigants would fit θ and β Canis Majoris, the brighter stars on each side of Sirius. The talents to be awarded could then become ξ^1 and ξ^2 Canis Majoris.

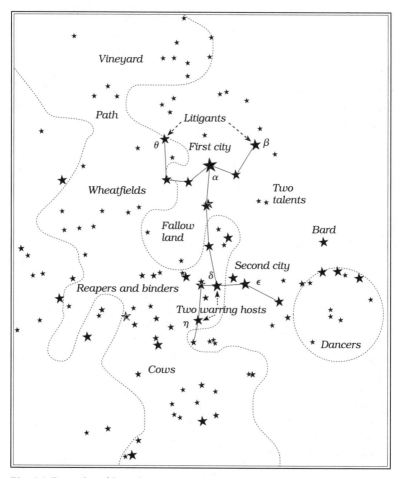

Fig. 56 *Examples of how descriptive narrative concerning the shield of Achilles could define stars and the Milky Way in the constellation of Canis Major. The Milky Way runs down the centre of the diagram, between the meandering dotted lines.*

About the other city there lay encamped two hosts in gleaming armour . . . But the men of the city . . . armed themselves for a surprise; their wives and little children kept guard upon the walls, and with them were the men who were past fighting through age; but the others sallied forth with Ares and Pallas Athene at their head – both of them wrought in gold and clad in golden raiment, great and fair with their armour as befitting gods, while they that followed were smaller. When they reached the place where they would lay their ambush, it was on a riverbed to which livestock of all kinds would come from far and near to water; here, then, they lay concealed, clad in full armour. Some way off them there were two scouts who were on the look-out for the coming of sheep or cattle, which presently came, followed by two shepherds who were playing on their pipes, and had not so much as a thought of danger. When those who were in ambush saw this, they cut off the flocks and herds and killed the shepherds. Meanwhile the besiegers, when they heard much noise among the cattle as they sat in council, sprang to their horses, and made with all speed towards them; when they reached them they set battle in array by the banks of the river, and the hosts aimed their bronze-shod spears at one another. (18.509)

This second city is ε Canis Majoris, the second brightest star in the constellation, and here Homer turns his attention to the southern part, or rear legs, of the constellation. The two hosts besieging the city are η and δ, stars on either side of ε Canis Majoris. Assorted groups such as wives, children and old men suggest clusters of fainter stars. The appearance of Ares and Athene outside the zodiac is surprising, but they may be being compared with Sirius and β Canis Majoris. In spite of Sirius now clearly being white, Ptolemy described the star as being 'reddish', while Mirzam, β Canis Majoris, is a white star. Homer in this passage may be comparing Sirius to Ares as the red planet Mars, and Mirzam to Athene as the planet Jupiter.

THE FALLOW FIELD

He wrought also a fair fallow field, large and thrice ploughed already. Many men were working at the plough within it, turning their oxen to and fro, furrow after furrow. Each time that they turned on reaching the headland a man would come up to them and give them a cup of wine, and they would go back to their furrows looking forward to the time when they should again reach the headland. The part that they had ploughed was dark behind them, so that the field, though it was of gold, still looked as if it were being ploughed – very curious to behold. He wrought also a field of harvest corn, and the reapers were reaping with sharp sickles in their hands. Swathe after swathe fell to the ground in a straight line behind them, and the binders bound them in bands of twisted straw. There were three binders, and behind them there were boys who gathered the cut corn in armfuls and kept on bringing them to be bound: among them all, the owner of the land stood by in silence and was glad. The servants were getting a meal ready under an oak, for they had sacrificed a great ox, and were busy cutting him up, while the women were making a porridge of much white barley for the labourers' dinner. (18.541)

Although the Milky Way is a continuous band surrounding the Earth, it has indentations of dark patches of sky in which there are few stars or even none at all. It is proposed that one of these dark patches is the fallow field. Homer says the field was dark behind them and yet still shone with gold – a curiosity that can be explained by the backcloth of the Milky Way illuminating the area of dark sky. The reapers and binders suggest brighter stars set against the background of the Milky Way (the field of harvest corn), while the man with the cup of wine is a bright star just outside its boundary.

The Vineyard

He wrought also a vineyard, golden and fair to see, and the vines were loaded with grapes. The bunches overhead were black, but the vines were trained on poles of silver. He ran a ditch of dark metal all round it, and fenced it with a fence of tin; there was only one path to it, and by this the vintagers went when they would gather the vintage. Youths and maidens all blithe and full of glee, carried the luscious fruit in plaited baskets; and with them there went a boy who made sweet music with his lyre, and sang the Linus-song with his clear boyish voice. (18.561)

The vineyard lies in the Milky Way to the north of Sirius at a place where dark skies (bunches of black grapes) lie among its 'silver poles'. The vineyard is reached by a path or a narrow part of the Milky Way. The dark ditch around the vineyard is the dark sky on each side of the Milky Way, which itself is bounded by a fence of tin.

HERD OF HORNED CATTLE

He wrought also a herd of horned cattle. He made the cows of gold and tin, and they lowed as they came full speed out of the yards to go and feed among the waving reeds that grow by the banks of the river. Along with the cattle there went four shepherds, all of them in gold, and their nine fleet dogs went with them. Two terrible lions had fastened on a bellowing bull that was with the foremost cows, and bellow as he might they haled him, while the dogs and men gave chase: the lions tore through the bull's thick hide and were gorging on his blood and bowels, but the herdsmen were afraid to do anything, and only hounded on their dogs; the dogs dared not fasten on the lions but stood by barking and keeping out of harm's way. (18.573)

Having described the bucolic scene of the vineyard in the north of the constellation, Homer focuses on stars in the southern regions of Canis Major. Cows of gold and tin grazing among waving reeds are very unlikely, but it is possible to see them as metaphors for yellow and white stars standing out against the background of the Milky Way. Homer reminds us of the overall view of Canis Major with the four golden shepherds representing four of the brightest stars of the constellation – α, β, ϵ and η – while the nine dogs become nine bright stars in the constellation outline.

Homer next focuses the observer's eye on a group of three bright stars through the image of two lions about to attack a bull. The bellowing bull is proposed as δ Canis Major and the two lions are η and ϵ, who later gorge themselves on the bull's entrails – the many stars surrounding δ Canis Majoris.

The Pasture

> The god wrought also a pasture in a fair mountain dell, and large flock of sheep, with a homestead and huts, and sheltered sheep folds. (18.588)

Description of the Milky Way now continues with an overall view of brighter groups of stars: 'a homestead', 'huts' and 'sheep folds' – shining on the starlit background of 'the large flock of sheep'.

> Furthermore he wrought a green, like that which Daedalus once made in Knossos for lovely Ariadne. Hereon there danced youths and maidens whom all would woo, with their hands on one another's wrists. The maidens wore robes of light linen, and the youths well-woven shirts that were slightly oiled. The girls were crowned with garlands, while the young men had daggers of gold that hung by silver baldrics; sometimes they would dance deftly in a ring with merry twinkling feet . . . and sometimes they would go all in line with one another, and much people was gathered joyously about the green. There was a bard also to sing to them and play his lyre, while two tumblers went about performing in the midst of them when the man struck-up with his tune. (18.590)

Youths and maidens, holding hands and dancing upon a green, take Homer to yet another part of the Milky Way, possibly the south-east corner, with fainter stars, close together, being prominent. The golden daggers and silver baldrics suggest yellow and white stars, while the bard may be an isolated single star.

> All round the outermost rim of the shield he set the mighty stream of the river Oceanus. (18.607)

The stream of Oceanus is the horizon (see Chapter 6, page 164).

Achilles' Armour

Compared with the detail lavished on the shield, at the end of Book 18 Homer spends little time on the armour of Achilles, but his astronomical imagery of all the stars of Canis Major is still rich:

> Then when he had fashioned the shield so great and strong, he [Hephaestus] made a breastplate also that shone brighter than fire. He made a helmet, close-fitting to the brow, and richly worked, with a golden plume overhanging it; and he made greaves also of beaten tin. (18.609)

In astronomical terms, the helmet is the first part of Canis Major to rise above the horizon, and its 'golden plume' is the Milky Way; the breastplate is brilliant Sirius in the heart or breast of the constellation.

With his constellation placed in the heavens, Achilles returns to the battlefield of Troy clad in brilliant new armour and carrying his marvellous shield. The great warrior will fulfil his destiny and ensure that the Trojans will not halt the movement of the heavens, and that Troy – and Ursa Major – will inevitably fall. All is not well for Achilles, however, and he is warned that the movement of the heavens that returned his personal star to the skies will also ensure his own downfall. His horse Xanthus warns him, 'Yet the hour of your death is drawing near; and it is not we who will be the cause of it, but a great god and the strong hand of destiny' (19.408). Homer uses the magical device of a talking horse to warn that the precession of the equinoxes – the hand of destiny – will ensure that Sirius will in the future once more vanish for thousands of years.

Precession and the Changing Equinoxes

In the *Iliad*, Homer tells of the movement – or precession – of the vernal and autumnal equinoxes through three 'generations' or constellations over a period of almost six thousand years. The vernal equinox was in Cancer *c*. 8000 BC and moved

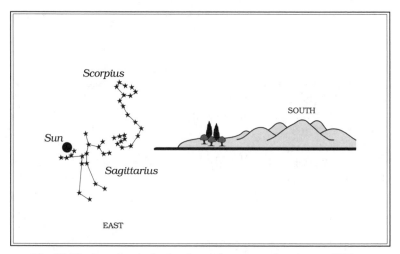

Fig. 57 The Sun rises in Sagittarius at the autumnal equinox c. 6500 BC.

to Gemini *c.* 6500 BC before precessing into Taurus *c.* 4400 BC, where it stayed until *c.* 2200 BC. During the same period the autumnal equinox was first in Capricorn before moving into Sagittarius (fig. 57) and then into Scorpius (fig. 58). Towards the end of the *Iliad*, Homer states that the vernal equinox will next move into Aries and the autumnal equinox into Libra (beginning at 2200 BC).

Harald Reiche makes the case for the movement of equinoxes being clearly observable in prehistoric times:

Consider the slow eastward slippage, past a fixed and ancient horizon marker, of the familiar constellation marking solstices or equinoxes, clearly noticeable after but a few generations and entailing the gradual obsolescence of any given polar star . . . Thus a constellation hitherto heralding sunrise at vernal equinox would seem to rise ever earlier with respect to sunrise until it had slipped so far eastward beneath the eastern horizon as to be still invisible at sunrise . . . meanwhile its replacement would, in horizon terms, descend from the sky to be the new harbinger of sunrise at vernal equinox.[4]

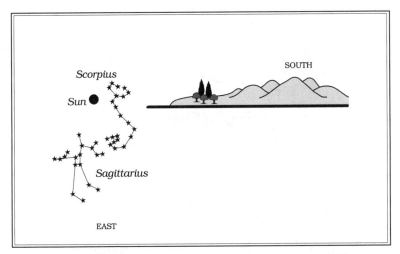

Fig. 58 *The Sun rises in Scorpius at the autumnal equinox c. 4400 BC.*

Changes in the heavens might not have been noticeable in one lifetime, but over three – from grandparent to grandchild – they could have been.

Homer begins his exposition of precession in Book 4 when the autumnal equinox is moving from the 'Trojan' constellation of Sagittarius into the 'Greek' constellation of Scorpius. At the same time, the vernal equinox is moving from Trojan Gemini to the Greek allies in Taurus. As with other astronomical allegories in the *Iliad*, these events are made memorable by the amount of blood that is shed as Trojans kill Greeks to try to halt precession, and Greeks kill Trojans to ensure it continues. No epithet could be more apt than that of 'foolish' given to the Trojan ally Pandarus (fig. 59), who tried to stop the autumnal equinox moving from his own constellation of Sagittarius to Scorpius, the celestial home of Menelaus of Sparta (fig. 60). When Pandarus broke a truce and fired an arrow at the heart of Menelaus, it brought about a renewal of the fighting that culminated in the fall of Troy. Menelaus' personal star is Antares, α Scorpii, in the 'heart' of the constellation. If Menelaus had

209

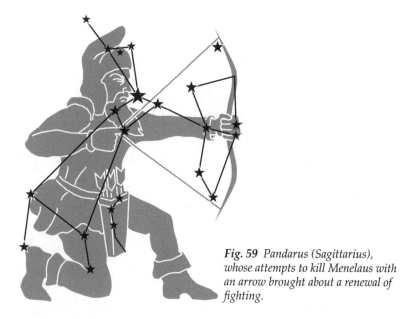

Fig. 59 *Pandarus (Sagittarius), whose attempts to kill Menelaus with an arrow brought about a renewal of fighting.*

been struck there he would have died and, in story at least, the precession of the equinoxes would have stopped – the *Iliad* would have come to a premature end with the lifting of the siege and Troy would have survived. The timely intervention of the goddess Athene ensured that this could not happen, by deflecting the arrow so that Menelaus was wounded in the 'belt' (4.127), a star lower in the body of the constellation. It was about 4400 BC when the celestial events represented by this squabble occurred and the autumnal equinox precessed into Scorpius. Within a short time Pandarus had been killed in battle, and with Sagittarius then 'dead' the autumnal equinox precessed without hindrance.

To make sure this important lesson in precession was not forgotten, Homer repeated it in different parts of the *Iliad*, on each occasion in a different way. In the zodiac on the opposite side of the sky from Sagittarius, and connected by the ribbon of light of the Milky Way, lies the Trojan constellation of Gemini (the

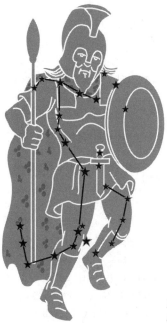

Fig. 60 *Menelaus, the object of Pandarus' assassination attempt.*

Lycians) and the Greek stronghold of Taurus (the Cretans). Just as Sagittarius (Pandarus) had to die so that precession of the autumnal equinox would take place, Gemini also has to 'die' so that the vernal equinox can move from Gemini to Taurus. But Homer takes much more time to describe the demise of Gemini than that of Sagittarius. In a number of actions, most of Gemini's Lycian warrior-stars are killed before their leader, Sarpedon, is lanced near the heart by Patroclus in Book 16. Homer reaffirms the inexorable nature of precession when Zeus, the immortal father of Sarpedon and controller of the heavens, is persuaded by Hera not to save his son from death (16.440).

Homer compares the changing of the seasons with changes in the heavens in some well-known words of Glaucus, joint commander with Sarpedon in the passing vernal constellation of Gemini: 'Men come and go as leaves year by year upon the trees. Those of autumn the wind sheds upon the ground, but when spring returns the forest buds forth with fresh vines. Even so is it with the generations of mankind, the new spring up as the old are passing away' (6.145).

To describe the eventual movement of the vernal equinox from Taurus to Aries *c.* 2200 BC, Homer tells a long and complicated story of the bravery of Meleager (9.527). For the change of the autumnal equinox from Scorpius to Libra at the same time, Homer says in Book 18 that Polydamas (who shares the star α^2 Librae with Helen of Troy) can 'look both forwards and backwards' (18.250), or see the changes wrought by precession both in the past and in the future.

Nestor is another wise old warrior with knowledge of precession. Homer says 'two generations of men born and bred in Pylos had passed away under his rule, and he was now reigning over the third' (1.250), which classicists have said would make Nestor about seventy years old. But the generations about which Homer speaks are astronomical, not those of mortal lives. From Nestor's position in the ancient constellation of Auriga above the zodiac, he has watched the vernal equinox precess over three 'generations' from Cancer to Gemini and on to the third in Taurus (fig. 61).

The Changing Pole Stars

In the course of a few hours sky-watching at night, the stars and constellations seem to be rotating around a particular point in the heavens. In the northern hemisphere today that point, where an extension of the Earth's axis of rotation would meet the celestial sphere, is close to Polaris in the constellation of Ursa Minor. It will not always be Polaris that indicates the north celestial pole, however. As the Earth rotates on its axis each day during its yearly journey around the Sun, the gravitational pull of the Sun and Moon make it wobble slowly as it spins. The result is that each end of the extended axis of rotation will slowly describe a circle on the celestial sphere,

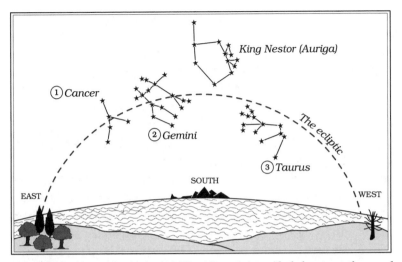

Fig. 61 *From his position in Auriga, King Nestor is in an ideal place to see the vernal equinox pass over 'three generations' from Cancer to Gemini and on to Taurus.*

pointing to different stars in a cycle that lasts almost 26,000 years (see fig. 6). So, as the centuries pass, the Earth's extended axis will point not towards Polaris but instead towards γ Cephei, the next significant star to become pole star.

The north pole star was important for travel and navigation in both ancient and much more modern times. Homer knew a great deal about the changing pole stars, and events in the *Iliad* concern the Greek Bronze Age, when the north pole star was Thuban, α Draconis. The warrior associated with that star was Arcesilaus, a commander of the Boeotians, whose name means Leader of Men – and this he certainly was, for when he reigned as pole star between about 4400 BC and 1800 BC he was a guide for travellers and seafarers.

As the north celestial pole moved closer to Thuban in Draco, so too did the Trojan constellation of Ursa Major rise higher in the sky. But Draco is not a very dramatic sight in the night sky, and when Thuban reigned it was Ursa Major – and the Big Dipper in particular – that dominated the skies near the north celestial pole.

The fall of Troy is an allegory for the decline of Ursa Major in the sky. The Trojans fought valiantly, but in vain, to kill all past and present pole stars and so stop precession and maintain the status quo. Inevitably, their struggle was doomed, and, as the north celestial pole began to move further away from Thuban, stars in Ursa Major that had once never set dipped closer to the horizon. In our times a small number of stars in Ursa Major set each day, and more of its stars will set below the horizon as precession continues. In this sense Troy is still falling. Hector, the wise Trojan warrior, knew that his army could stop neither the continual precession of the equinoxes nor the changing of the pole stars, and Homer has him invoke an image of the Earth's extended axis – the spear shaft of the heavens – when he says, 'Deep in my heart I know the day is coming when holy Ilium will be destroyed, with Priam and the people of Priam with the good ashen spear' (6.446). The imaginary shaft which points far out to celestial north belongs to the Greeks – King Agamemnon's sceptre (2.100), the great spear of Achilles (19.387) and the 22-cubit pole of Great Aias (15.678) are all metaphors for it – and its movement cannot be halted. The bitter fate of the Trojans and their allies is that the continued movement of 'the ashen spear' will be their downfall.

One interesting aspect of Edna Leigh's work on the pole stars concerns the deaths of two men by the name of Schedius of the Greek regiment from Phocis. First to die (15.515) is Schedius the son of Perimedes, 'leader of the Phocians', but it has been suggested that Homer made an error, inasmuch as in the Catalogue of Ships he had already named Schedius, son of mighty Iphitus, as commander of the same regiment. Edna Leigh looked at original texts and expressed a view that Homer was not 'nodding', but, by twice using the name Schedius for a leader of the Phocians, was drawing attention to the two stars from their home constellation of Cygnus that had in the distant past indicated celestial north. The astronomical status of these

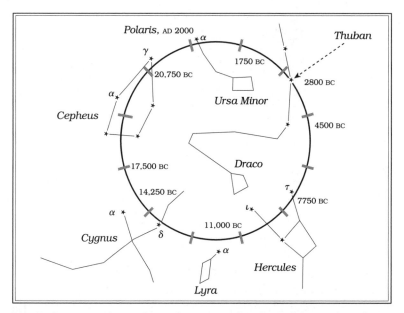

Fig. 62 *Stars on or close to the north precessional circle which have indicated celestial north. Today the pole star is Polaris in Ursa Minor, but in the course of time Polaris will be superseded by γ Cephei.*

men is indicated in that each is killed by powerful Hector. Schedius the son of Iphitus is struck in the collarbone at 17.306, and is allocated to Deneb, α Cygni. In Book 15, Homer does not say where Schedius the son of Perimedes was hit, but, as can be seen in fig. 62, he is allocated to δ Cygni, the other star in Cygnus that once led men towards the pole.

Fig. 62 shows the stars that have in the past and will again in the future indicate celestial north during the 26,000-year precessional cycle, and Table 11 lists the warriors that Homer identified with these stars. Stars shown in bold are pole stars directly on or almost on the northern precessional circle. Although the bright stars of Deneb and Vega lie some distance from the circle, they are sufficiently close to be indicators of celestial north.

215

Table 11 *Pole stars of the precessional cycle*

Star name	Catalogue	Warrior	Notes
	τ **Herculis**	Philoctetes	Bitten by a snake (2.722)
Edasich	ι Draconis	Prothoenor	Killed by Polydamas (14.450)
Thuban	α **Draconis**	Arcesilaus	Killed by Hector (15.329)
Kochab	β Ursae Minoris	Podarces	Survives (2.704)
Polaris	α **Ursae Minoris**	Protesilaus	First warrior killed in the Trojan War (2.698)
Alrai	γ **Cephei**	Medon	Killed by Aeneas (15.332)
Alderamin	α **Cephei**	Little Aias	Survives (2.527)
Deneb	α Cygni	Schedius	Killed by Hector (15.515)
	δ Cygni	Schedius	Killed by Hector (17.306)
Vega	α Lyrae	Eurypylus	Wounded by Paris (11.580)
	ι Herculis	Medon	Killed by Aeneas (15.332)
	τ **Herculis**	Philoctetes	Reappears at *Odyssey* 3.189

Hercules, one of the most acclaimed Greek heroes, is referred to in the *Iliad* only when his name is invoked as that of a warrior of mighty deeds. Philoctetes, the archer, takes his place in the constellation of Hercules because he was, according to legend, awarded Hercules' bow as a reward for setting light to Hercules' funeral pyre and releasing the hero from his agonizing life. From the Catalogue of Ships we learn that Philoctetes is bitten in the foot and takes no part in the *Iliad*, but he does return in the *Odyssey*. His recovery indicates that Homer was aware, just as were the Egyptians, that the circle of precession repeats itself. The star τ Herculis, Philoctetes' wounded foot, was a pole star in the past and will return to be a pole star in the future. The name of the Greek warrior Medon appears twice in the table, and his role both in the *Iliad* and in Homeric astronomy is complex.

Homer and Time

'Homer's war, the war of the poems and of the tradition, is a timeless event floating in a timeless world,' wrote the scholar Moses Finley.[5] Homeric astronomy is also timeless, in the sense that it is concerned with phenomena that will continue until our solar system is no more. Even the time span of those celestial events recorded in the *Iliad* can hardly be calculated in terms of human generations. When we searched the *Iliad* for astronomical knowledge, one of our most exciting experiences was the growing revelation of Homer's awareness of events that were separated by thousands of years. When the idea first arose that Homer was describing the reappearance of Sirius *c.* 8900 BC, it was a severe test for the imagination, but repeated analysis did not cause us to abandon the hypothesis.

Computer programs showed that significant astronomical occurrences in the *Iliad* began as long ago as the ninth millennium and ended at about 1800 BC. During this period the vernal and autumnal equinoxes each moved through three constellations – as, of course, did the summer and winter solstices. In such allegories as the attempt by Pandarus to kill Menelaus, the killing of two sons of Antenor by Agamemnon, and the death of Sarpedon, Homer highlighted an equinox or solstice in each of the twelve signs of the zodiac.

Because it is not known how the ancients determined the boundaries of the zodiacal constellations, it is not possible to put a precise date to the precession of the spring and autumn equinoxes, but Table 12 may be useful as a guide to the timescale of Homeric epic. Each constellation has the name of the commander of the regiment it represents in parentheses after it.

Table 12 Homer's equinoxes[6]

Constellation	Date of entry	Duration
Autumnal equinoxes		
Capricornus (Peiros)	8000 BC	1,500 years
Sagittarius (Pandarus)	6500 BC	2,100 years
Scorpius (Menelaus)	4400 BC	2,600 years
Vernal equinoxes		
Cancer (Diores)	8000 BC	1,500 years
Gemini (Sarpedon)	6500 BC	2,100 years
Taurus (Idomeneus)	4400 BC	2,600 years

In terms of the *Iliad*, these dates signify that the confrontations over the movement of the autumnal equinox between Pandarus (Sagittarius) and Menelaus (Scorpius) (4.93) represent astronomical events that would have taken place *c*. 4400 BC. It follows that the movement of the vernal equinox from Gemini to Taurus with the death of Sarpedon (16.481) would have been at the same period.

Edna Leigh declared that, on an annual basis, the events of the *Iliad* took place in the weeks before and after a summer solstice long ago, and such a time is indicated when Homer says in Book 18, two days before Hector is killed, that 'the Sun was disposed to linger but at last he set' (18.239) – in other words, the hours of daylight were at their longest. In an allegory at the beginning of Book 2, Agamemnon (Regulus, α Leonis) rises from his bed at dawn and is dressed in such finery that he can be compared with the Sun rising at the same time and covering his personal star (2.41). Such an event could have occurred at a summer solstice at about 5500 BC, but more precise calculations than are available to the authors may give a more accurate date for Agamemnon's day of glory on Midsummer's Day.

There must be speculation about how Homer knew of events which occurred long before his time. It has been shown that he predicted a cycle which would lead to ι Herculis becoming a pole star. To claim that knowledge of an event that happened *c*. 11,000 BC could have found its way to Homer through the millennia by word of mouth raises many questions. On the other hand, to propose that this event could have been deduced by ancient peoples would add another dimension to their achievements. The return to the skies of Sirius *c*. 8900 BC might have been deduced from a knowledge of the movement of the heavens, but if ever an astronomical occurrence were to be preserved in epic then it would have been this arrival above the horizon of the brightest star. The precessional sequence of equinoxes and the return of Sirius/Achilles concern events thousands of years before Homer, but an equally important

218

element involving time – the fall of Troy – took place rather closer to his period. Thuban declined as pole star from about 2800 BC to around 1800 BC. This indicates that peoples long before Homer had a sophisticated knowledge of astronomy, but raises the question of why Homer did not describe the skies of his own era. A glance at a star chart provides an answer. When the *Iliad* was probably composed, *c*. 745 BC, the Earth's axis was pointing to a part of the precessional circle with no exceptional stars. The next time a star as prominent as, or brighter than, Thuban indicated celestial north was from *c*. 1100 AD, when the pole shifted towards Polaris in Ursa Minor, the present pole star, with a magnitude of 2.1.

Table 13 (*overleaf*) relates incidents in Homer's narrative to the dates of the astronomical events which they portray.

To grasp the time span of astronomical occurrences in the *Iliad* required a considerable mental adjustment, and with it came an awareness that the theme or plot of the *Iliad* was entirely controlled by Homer's aim to preserve knowledge of the effects of precession. Edna Leigh recognized that the decline of the star Thuban was synonymous with the fall of Troy, and saw the significance of the killing of warriors whose stars lay on the precessional circle. When the departure and return of Achilles was added to these, the three allegories of precession were complete.

The Poet-Singers and the 'Memory Wheel' of the Zodiac

It was seen in Chapter 2 how techniques to aid memory could have helped Homeric bards to recall large amounts of data from memory. Fig. 63 shows how a 'memory wheel' based on the zodiac could have made it possible to commit to memory the sequence in which the equinoxes precessed from one division of the zodiac to another over thousands of years.

The zodiac is laid out in the form of a wheel, and assigned to each of the twelve zodiacal constellations is a Greek or Trojan

Table 13 *Calendar of Homeric astronomy*

Incident	Date
1. Homer refers to the deeds of heroic Hercules at a time long before the earthly events of the *Iliad*. This suggests an older astronomical order when the stars ι and τ Herculis in the constellation Hercules indicated celestial north.	ι Herculis = *c*. 11,000 – 9000 BC τ Herculis = *c*. 9000–7000 BC
2. Achilles/Sirius returns to the skies of Greece.	*c*. 8900 BC
3. The beginning of Homer's exposition of the precession of the vernal and autumnal equinoxes and the summer and winter solstices.	*c*. 8000 BC
The equinoxes and solstices pass into the 'second generation' of constellations.	*c*. 6500 BC
4. Agamemnon dresses for battle – the Sun rising in Leo over α Leonis at midsummer.	Summer solstice, *c*. 5500 BC.
5. Pandarus attacks Menelaus – an allegory for the passing of the autumnal equinox from Sagittarius to Scorpius.	*c*. 4400 BC
6. As the autumnal equinox changes, the vernal equinox passes from Gemini to Taurus when Patroclus kills Sarpedon.	*c*. 4400 BC
7. Homer forecasts that the autumnal equinox will pass from Scorpius to Libra and the vernal equinox from Taurus to Aries.	*c*. 2200 BC
8. The fall of Troy is an allegory for the decline of Thuban as pole star.	*c*. 2800–1800 BC
9. Homer records a rare conjunction of planets.	Dawn on 5 March 1953 BC
10. Homer wrote the *Iliad*.	*c*. 745 BC
The time of year when the *Iliad* takes place:	The weeks before and after a summer solstice
The climax of the Iliad is the longest day, when 'Hera sent the busy Sun, loth though he was, into . . . Oceanus' (18.240).	Midsummer's Day

regiment. The wheel shows how the equinoxes and the solstices precessed *c*. 4400 BC – a period of precession which Homer treats in considerable detail in the *Iliad*, beginning, as we have seen, when foolish Pandarus tries to halt precession by attacking Menelaus, an allegory for when Scorpius ousted Sagittarius as the constellation of the autumnal equinox. At the same time, the vernal equinox moved from Gemini to Taurus, the summer solstice from Virgo to Leo, and the winter solstice from Pisces to Aquarius.

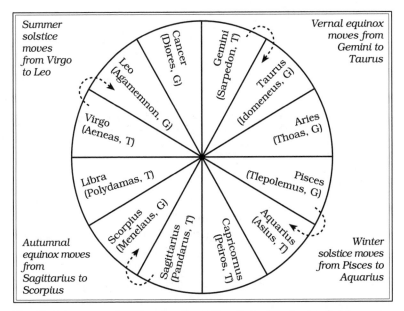

Fig. 63 *A 'memory wheel' based on the zodiac. Each of the twelve divisions represents a Greek or Trojan regiment, indicated here by the name of its commander and the letters G or T, for Greek or Trojan.*

From the wheel, it can be easily deduced that the next period of precession would take the autumnal equinox from Scorpius to Libra, the vernal equinox from Taurus to Aries, the summer solstice from Leo to Cancer, and the winter solstice from Aquarius to Capricornus. Equally easily, it could have been brought to mind that the previous sequence of precession took the autumnal equinox from Capricornus to Sagittarius, the vernal equinox from Cancer to Gemini, the summer solstice from Libra to Virgo, and the winter solstice from Aries to Pisces.

Whether recalling such information from memory or even using the night sky as an aide-mémoire, ancient people could thus have imprinted on their minds the sequence and time-scale of the precession of the equinoxes over 'three generations' or more than six thousand years.

8

Homer's Earth-Centred Universe

. . . there is one philosophical problem in
which all thinking men are interested. It is
the problem of cosmology: *the problem of
understanding the world* . . .

Karl Popper, *The Logic of Scientific Discovery*
(1959)

The quest to discover the nature of the universe and the
place of the Earth within it has been pursued by mankind
since earliest times. Modern scientists evolve theories about
Big Bangs and black holes, and spacecraft journey to the Moon
and planets to seek a greater understanding of the cosmos. In
the ages before Homer, and during the flowering of Greek
culture in the centuries afterwards, philosopher-scientists pondered the same questions and displayed a remarkable unity of
thought.

It is not known from where or when originated Homer's
notion that the Earth stood isolated in space at the centre of a
celestial sphere on which the stars appeared to be fixed and
which rotated once each day around us. This concept of a geocentric universe remained predominant in astronomical
thought in later centuries and, considerably enhanced by the
application of geometry, influenced Western astronomy until

the sixteenth century AD. It was then that Copernicus showed that Ptolemy and Homer and generations of other ancient scientists were mistaken, and that the Sun is at the centre of our particular system, with the Earth and other planets orbiting around it.

Homer's ingenuity in recording his theory in narrative is remarkable, and he enhances the basic idea of an Earth-centred universe by establishing the concepts of the zenith and the nadir of the celestial sphere as well as the position of the north celestial pole. The zenith is the highest point of the celestial sphere above an observer at any place on Earth. The nadir is the lowest point of the celestial sphere below an observer at any place on Earth. Only half of the celestial sphere can be seen at any one time, the other part being below the horizon.

Homer begins the construction of his universe with allegories that delineate the two halves of the celestial sphere. In so doing he projects two views of the skies that include all the stars and constellations that would pass across the skies of Greece during one complete rotation of the celestial sphere. He then shows how the celestial sphere rotates around the earth, and concludes his exposition by placing in position the north celestial pole, the zenith and the nadir. Episodes in which each of these component parts is defined in the Books of the *Iliad* are as follows:

1. Scorpius rising in the east as Orion sets in the west gives a view of half of the celestial sphere. This is portrayed in the confrontations between Menelaus and Paris in Book 3. See fig. 64.
2. The apparent rotation of the celestial sphere is shown when the constellation of Libra eventually follows Orion and sets below the horizon. This incident also takes place in Book 3, when Helen (α^2 Librae) reluctantly joins her husband, Paris, in bed (below the horizon) after the passage of some time. See fig. 65.

3. The second half of the celestial sphere is defined by the rising of Perseus and the setting of Virgo. This is portrayed in Book 5, when Pandarus (Sagittarius) and Aeneas (Virgo) set out to kill Diomedes (Perseus). See fig. 66.

4. Homer seeks to confirm the rotation of the celestial sphere in Book 6, when Hector is sent on a journey which represents Orion's apparent path around the Earth. See fig. 67. This important message is repeated in Book 24, when Achilles has a troubled night's sleep. See fig. 69.

5. Homer's obsession with the celestial sphere and the curvature of the heavens is reinforced in a brief but effective incident when he describes the changing attitude of Orion as it crosses the skies. This occurs in Book 6, when Paris returns to the fighting and tells his brother that he should go first but eventually he will overtake him. See fig. 68.

6. That the Earth lies suspended in space at the centre of the universe and the positions of the zenith, the nadir and the north celestial pole on the celestial sphere are established when Zeus reads the Riot Act to the gods in Book 8. See fig. 70.

First Half of the Celestial Sphere

Homer delineates half of the celestial sphere in Book 3, when Paris, the Trojan seducer of Helen, challenges any of the Greek warriors to meet him in single combat to decide the outcome of the war. Paris soon flees in terror when Menelaus, the former husband of Helen, picks up the gauntlet and strides on to the battlefield to meet him. 'Paris quailed as he saw Menelaus come forward, and shrank in fear of his life under cover of his men . . . even so did Paris plunge into the throng of Trojan warriors, terror-stricken at the sight of the son of Atreus' (3.30). Paris, believing honour and valour to be secondary to self-preservation, fades away from the scene.

To make it easier for audiences in Homer's times to follow the astronomical events in the sky, the protagonists are shown in the roles of the constellations they represent, rather than as individual stars, Paris being Orion, and Menelaus being Scorpius.

Realizing the potential of this confrontation was one of the first advances we made in the interpretation of the astronomical content of the *Iliad*. The breakthrough came when we examined a computer program to see if there was a reason why Homer had used the men who represent Scorpius and Orion for this event. It was immediately apparent that this allegory defined a view of half of the entire celestial sphere, from Scorpius to Gemini and Orion. It became clear, too, that this area of sky was the battlefield of Troy on which most of the celestial events described in the *Iliad* take place. The significance of Paris (Orion) fleeing from the clutches of Menelaus (Scorpius) also became apparent: Homer was introducing the idea of a rotating celestial sphere, and was describing how, as Scorpius rose, Orion would very soon dip below the horizon – or how Paris would flee below the horizon where he could not be seen by Menelaus (fig. 64). So important does Homer consider this wide view of the celestial sphere between Scorpius and Orion that he stages another encounter between Paris and Menelaus, with additional astronomical detail (3.340). On this occasion the two men fight, but as Menelaus gains the upper hand Paris is spirited away to his bedchamber by the goddess Aphrodite as Orion sets.

Helen Joins her Husband

Below the horizon, Paris awaits the arrival of Helen of Troy, who has not been at all impressed with his lack of courage and is reluctant to join him (3.410). It is not surprising that Helen hesitates, for her personal star α^2 Librae is in the sky for several hours after Orion has set (fig. 65). Only after Homer has had Aphrodite threaten Helen with dire penalties is she reunited 'in bed' with Paris below the horizon.

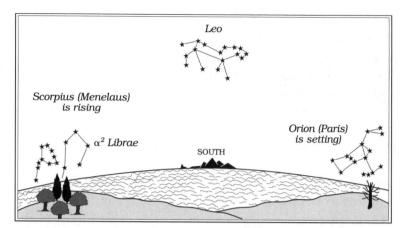

Fig. 64 In their first encounter, Menelaus (Scorpius) and Paris (Orion) set the boundaries for the first of Homer's two views of the celestial sphere. For clarity, all other constellations in this area of the sky, apart from Leo, have been omitted. This panorama of the heavens is the one in which most of the fighting in the celestial Siege of Troy takes place.

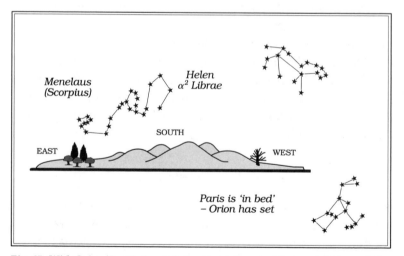

Fig. 65 With Orion (Paris) already below the horizon, and the constellations of Libra (Helen) and Scorpius (Menelaus) moving towards the western horizon, Homer shows how constellations move across the dome of the sky before setting in the west.

226

Meanwhile, what about Menelaus, whose constellation of Scorpius is still visible? With Paris safe in hiding with Helen, Homer says, 'Thus they laid themselves on the bed together; but [Menelaus] strode among the throng, looking everywhere for Paris, and no man, neither of the Trojans nor of the allies, could find him' (3.448). Menelaus could never find him, of course, because Paris was below the horizon, making love to the beautiful Helen. Boastful to the end, Paris tells Helen that if Menelaus and he meet again then he himself may be victor (3.440). That the two warriors will appear in the skies again is as certain as anything can be, for each day Scorpius and Orion are at opposite ends of the horizon for a brief time before Orion sets.

The Other Half of the Celestial Sphere

With the first half of the celestial sphere established, the question was raised, How did Homer describe the second half of the celestial sphere? Constellations in this view would be those which were below the horizon when Homer described the duels of Paris and Menelaus. The second view would include constellations in the zodiac from Taurus, Aries, Pisces, Aquarius, Capricornus and finally Sagittarius. However, the constellations in this second panorama of the skies lack the visual impact of those in the first. There are relatively few bright stars, and, apart from Sagittarius, the constellations are not so easily recognizable. A number of sequences in which Homer describes combat between Greek and Trojan opponents were run through on a computer, and when the strange death of Pandarus was examined it provided an answer.

Pandarus, the archer ally of Troy, whose home is in Sagittarius, had already been labelled 'foolish' for his ultimately unsuccessful attempt to stop the precession of the equinoxes by killing Menelaus (see Chapter 7). Having failed, Pandarus had to die, and, showing a remarkable economy of expression, Homer in Book 5 not only ties up loose ends of precession, but also depicts the other half of the celestial sphere. In narrative terms, this occurs when Pandarus and Aeneas (Virgo) mount a chariot to seek out Diomedes (Perseus) on the field of battle (5.220). Pandarus again fails to kill a foe when his spear strikes the shield of Diomedes, who then retaliates and kills his attacker. Aeneas stands over the body of his dead ally before being struck on the hip by a stone hurled by Diomedes and then taken to safety by the goddess Aphrodite.

In setting the boundaries of the second half of the celestial sphere, Homer tells us that Diomedes' 'limbs are as yet unwearied' (5.253), placing newly risen Perseus on the eastern horizon, and showing that the prominent long legs of the constellation would have been fresh for their journey across the heavens. The boundary on the western horizon is set by Virgo, which is about to set and will be followed by Sagittarius. This view of the celestial sphere overlaps with the first hemisphere in the same way that the edges of pages of modern atlases and road maps contain detail duplicated on pages showing adjacent areas.

When the opponents clash, Diomedes quickly turns the tables on the Trojans, and kills Pandarus in a manner that has long puzzled classicists but for which there is an astronomical explanation. When Diomedes casts his fatal spear, it enters Pandarus' head from above, strikes the top of his nose, passes through his mouth and exits under his chin. A simple enough description, perhaps, but hardly likely – because Pandarus is standing in a chariot high above Diomedes, who is fighting on foot. No truly satisfactory literary reason has been given to explain how Diomedes could have cast his spear high enough for it to arc through the air and hit Pandarus on the top of the head. The astronomical explanation is much more reasonable: with Diomedes' constellation of Perseus rising in the east and Sagittarius in the south-west, a spear launched by the Greek would follow the curvature of the heavens to the zenith (see fig. 66) before plunging down from the skies into the 'head' of

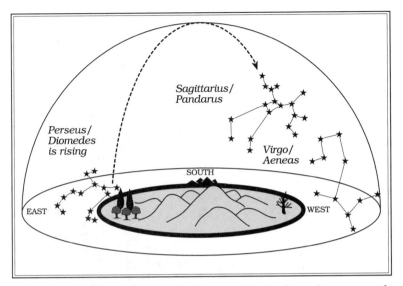

Fig. 66 *A spear thrown by Diomedes (Perseus rising in the east) arcs across the dome of the heavens to kill Pandarus, as Homer delineates his second view of the sky. The constellation Virgo, home of Aeneas, who accompanied Pandarus on his futile expedition, sets in the western horizon.*

Pandarus – or Sagittarius. The demise of Pandarus completes Homer's image of the celestial sphere.

The Revolving Celestial Sphere

As the Sun, Moon and stars rise and set on their daily journeys across the sky, they give the impression that the entire celestial sphere is rotating. To emphasize this idea, Homer sends Hector on a journey which represents that of Orion around the Earth, describing the setting of Orion in the west, its journey beneath the Earth, and its rising in the east before travelling again across the visible sky during one rotation of the celestial sphere (fig. 67).

In Book 6 Hector leaves the battlefield, visits the women of Troy, and says farewell to his wife and son for the last time, before returning to the battlefield with his brother Paris. At

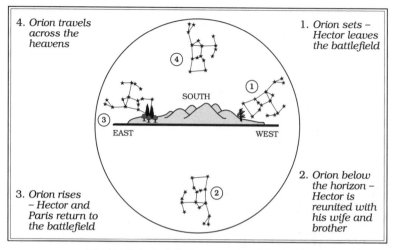

Fig. 67 Orion's daily journey around the Earth.

first Hector cannot find his wife, Andromache (β Leporis), and their son, Astyanax (λ Leporis), whose celestial home is the constellation Lepus, which lies at the foot of Orion and has long since set. Only when Hector leaves the daytime skies and goes below the horizon is he reunited with them. Hector also has the task of finding his brother Paris, still skulking below the horizon, and persuading him to return to the fighting.

THE CHANGING ATTITUDE OF CONSTELLATIONS MOVING ACROSS
THE SKY

After making his farewells to his family, Hector persuades his brother to return to the battle. Paris agrees, and tells Hector, 'Go first and I will follow. I shall be sure to overtake you' (6.341). As fig. 68 illustrates, this curious remark shows how constellations arc across the dome of the sky. When Orion first rises, and the constellation appears to stride across the sky, Hector's personal star, Rigel, β Orionis, leads the way or precedes that of Paris, Betelgeuse, α Orionis. When the constellation reaches its highest point, its changing attitude to the horizon shows that

230

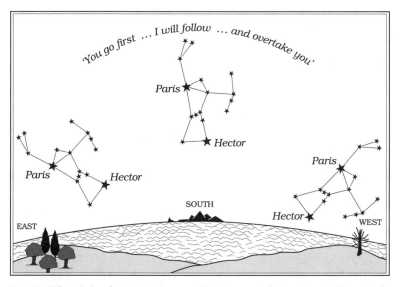

Fig. 68 *When Orion first rises, Hector, β Orionis, precedes Paris, α Orionis. As the constellation crosses the sky its attitude changes, and by the time Orion sets Paris has 'overtaken' Hector.*

Betelgeuse has almost 'caught up' with Rigel. Finally, as the constellation is about to set, Betelgeuse has 'overtaken' Rigel: Paris, who was once behind, is now in front.

This is not the only time Homer sends a constellation on a proving trip around the Earth. In Book 24, Achilles has a disturbed sleep in which 'he lay now on his side, now on his back, and now face downwards, till at last he rose and went out as one distraught, to wander upon the seashore' (24.10). We thought these words described Canis Major setting and journeying below the horizon before finally rising, but encountered a difficulty because literary translations of Achilles' restless sleep did not accord with the attitudes of Canis Major around the Earth. The mystery was resolved by turning to a literal translation of the Greek text: 'Remembering this [the death of Patroclus] a fresh tear he shed, sometimes on his sides lying down, sometimes (but) on the other hand backwards,

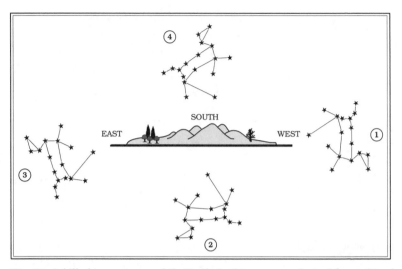

Fig. 69 *Achilles' journey around the Earth. In this sequence, derived from a literal translation of the Greek text, Achilles as Canis Major is (1) 'on his sides'; (2) 'backwards', or moving in the opposite motion to when in the visible sky; (3) 'head uppermost' when about to rise; (4) 'straight standing', or following its usual course across the sky. The shore of the 'salt sea' is the edge of the Milky Way.*

sometimes (but/and) head uppermost. Then straight standing he made circles deeply moved along the shore of the salt (sea).'[1] Although hardly as lyrical as the usual literary translations, the Greek text was astronomically more accurate and put the changing attitude of Canis Major into the correct sequence – see fig. 69. The seashore is one of a number of codewords Homer uses to describe the Milky Way, against which the constellation of Canis Major (Achilles) appears to stand.

Another excellent sequence showing how constellations arc across the sky is in the death of Hector (see Chapter 3, pages 72–7).

Homer Completes the Celestial Sphere

The incidents from the *Iliad* discussed earlier show Homer's understanding of the celestial sphere, and this understanding suggests that he would have known that stars were in the sky during the day even though they could not be seen because of the light of the Sun. From this would have followed the conclusion that stars seen at midnight at one time of year would at other times either be in the daytime skies or be visible around dawn or dusk. Even today constellations are often referred to as summer or winter and spring or autumn constellations visible in the night sky.

Interpretation of the *Iliad* strongly indicates that most of the fighting at Troy took place in the daytime skies around a summer solstice long ago on a celestial battlefield bounded by Scorpius and Orion. Edna Leigh wrote:

> Because Homer often tells of the daylight adventures of stars, the greater part of the *Iliad* may at first seem difficult to visualize. Homer positions his stars as an observer would see them appear to move across the skies – if the stars were visible by day. This factor indicates a highly developed science that assumes the heavens to be a sphere. With a thorough knowledge of the appearance of the heavens and a detailed understanding of movements of heavenly bodies at our command, we shall experience no more trouble in viewing Homer's celestial scenes via the Greek poet's epical images than did the people of antiquity.

That he was aware that constellations in the sky around midday at midsummer could be seen in the middle of the night at midwinter is explored in an astronomical interpretation of a raid on the Trojan lines by Odysseus and Diomedes during the hours of darkness (10.272). During this escapade, known as the Doloneia, not only are the constellations of Odysseus and Diomedes (Boötes and Perseus) in the sky, so too are those of Nestor (Auriga), who devised the raid, and Agamemnon (Leo), who agreed to it. These constellations had been in the daytime sky at the summer solstice, and are among those included in the view of the heavens fixed by the rising of Scorpius and the setting of Orion during the confrontations between Menelaus and Paris in Book 3. By setting Book 10 at night, Homer is indicating that the events it describes take place at the winter solstice. As the episode continues, he goes on to show how the same constellations which are in the noon sky at the summer solstice and in the midnight skies at the winter solstice can also be seen at sunset at the vernal equinox and at sunrise at the autumnal equinox. The night raid and other events that occur in Book 10 add a dimension to Homeric astronomy that will make a lengthy study in itself.

'Whoever was the first to maintain that the sky is studded with stars by day as well as by night, and that we fail to see them because they are obscured by the light of the Sun, was certainly a brilliant and courageous logician,' wrote André Danjon.[2] Homer certainly knew of this idea and its implications for the skies at different seasons of the year, but whether he was the first to deduce it or whether it was already ancient knowledge in his times may never be known.

The Earth Suspended at the Centre of the Celestial Sphere

To complete his picture of an Earth-centred universe, Homer fittingly uses the words of Zeus, commander of the heavens, to describe the Earth suspended in space, and to add the final embellishments of the zenith, the nadir and the north celestial pole to the celestial sphere.

Homer's exposition of an Earth-centred Universe begins in Book 8 with Zeus sitting on Mount Olympus and reading the Riot Act to his fellow gods. Under the threat of dire punishment, he orders them to stop taking sides in the war between the Greek and Trojans:

Hear me, said Zeus . . . If I see anyone acting apart and helping either Trojans or [Greeks], he shall be beaten inordinately ere he come back again to Olympus; or I will hurl him down into dark

Tartarus far into the deepest pit under the Earth, where the gates are iron and the floor bronze, as far beneath Hades as Heaven is high above the Earth, that you may learn how much the mightiest I am among you. Try me and find out for yourselves. Hang me a golden chain from Heaven, and lay hold of it all of you, gods and goddesses together – tug as you will, you will not drag Zeus the supreme counsellor from Heaven to Earth; but were I to pull at it myself I should draw you up with Earth and sea into the bargain, then would I bind the chain about some pinnacle of Olympus and leave you all dangling in mid firmament. (8.5)

Homer has identified Olympus, Zeus' home, as the zenith, the highest point on the celestial sphere above a person's head. The nadir, the lowest point of the celestial sphere directly beneath an observer at any point on Earth, is Tartarus, 'as far beneath Hades as Heaven [Olympus] is high above the Earth'. That the Earth is at the centre of the universe and isolated in space is determined when Zeus warns the gods that he could 'draw you up with Earth and sea . . . then would I bind [a golden] chain about some pinnacle of Olympus and leave you all dangling in mid firmament'.

Still missing from the sphere is the significant astronomical reference point of the north celestial pole. To show his awareness of this, Homer sends Zeus to his second home on Mount Ida:

Zeus lashed his horses and they flew forward nothing loth midway twixt Earth and starry Heaven. After a while he reached many-fountained Ida . . . he took his seat all glorious upon the topmost crests, looking down upon the city of Troy and the ships of the [Greeks]. (8.45)

Mount Ida is the north celestial pole, the point in the sky marked by the pole star, lying midway between Zeus' home on Olympus (the zenith) and the horizon. It is very significant that from Mount Ida Zeus can look down upon the city of Troy, for the pole star of the *Iliad* was Thuban, in the constellation of

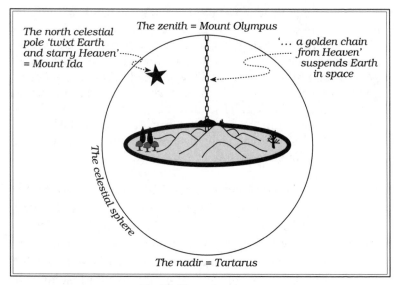

Fig. 70 Homer's universe.

Draco, which directly 'overlooks' Ursa Major, the celestial citadel of Troy. When viewed from Athens the pole star lies 38° above the horizon – close enough to 45° for it to be described as halfway between Earth and starry Heaven. How Homer's concepts of the Earth-centred universe are brought together can be seen in fig. 70.

The *Iliad* does not seem to indicate that Homer was aware that the Earth itself is a globe, and for this study it is assumed to be a flat disc as described by later Greek philosophers. This model helps to explain the idea first expressed in Chapter 6, page 164, that the mythological concept of the god Oceanus being a 'river' surrounding the Earth is a metaphor for the horizon. In view of Homer's considerable knowledge of astronomical and geographical matters, it would be strange if in this one instance he reverted to a fantasy explanation of a natural observation. From as long ago as Minoan Crete, peoples of the Aegean had travelled widely in the pursuit of trade, and it must have been obvious to them that, no matter how far they journeyed, they never arrived at this 'river'.

Another question raised by Homer's notion of the celestial sphere is, How did he cater for the movement of the Sun, the Moon and the five planets visible to the naked eye? That is, did he envisage just one sphere studded with stars and on which the Sun, the Moon and the planets had a mysterious life of their own, or did he think the sphere had a number of independent layers? The latter idea is similar to the system proposed by Eudoxus (408–355 BC) and to that preserved in the *Almagest* of Ptolemy. Eudoxus' system was based on concentric spheres centred on the Earth and 'was for two thousand years instrumental in shaping philosophical views on the general form of the universe'.[3]

We have shown in Chapter 4 that, some nine hundred years after Homer, Ptolemy used a method of identifying stars which gave strikingly similar results to the Rule of Wounding. It is therefore not unreasonable to investigate the possibility that Eudoxus and other post-*Iliad* Greek astronomers were influenced by Homer's views on other astronomical matters.

The explication of Homer's model of the universe reveals a relatively simple but nevertheless intellectually advanced concept of the Earth suspended in space and surrounded by a rotating sphere. Matters become more interesting when Homer's knowledge of the occultation of stars by the Moon and planets and of the independent paths of planets is considered. In the *Iliad*, gods representing the Moon and planets can assume the characters of mortals (stars), and this is Homer's way of recording the phenomenon of a body closer to the Earth (i.e. the Moon or a planet) temporarily obscuring a star.

It could not have escaped Homer that, for one celestial object to obscure another, one object must be nearer to the Earth than the other. At the simplest , this would have given a universe of two spheres – one sphere moving at a uniform rate for the stars, and another for the Sun, the Moon and the planets. However,

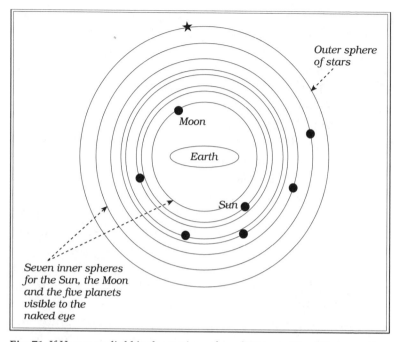

Fig. 71 *If Homer applied his observations of occultation and the differing motions of the planets to his concept of the universe, this diagram shows what might have been the result.*

Homer also records the journeys around the zodiac of the Moon and the visible planets. These objects quite obviously follow different paths and travel at different apparent speeds, which would make it impossible for them to be fixed to the same sphere. Did Homer then allocate a separate 'sphere' to the stars, the Sun, the Moon and each planet (fig. 71), and so view the universe in a similar manner to that of later Greek astronomers? Perhaps the idea of a universe constructed of separate spheres was already old before being enhanced by the mathematical and geographical propositions of Homer's successors and preserved in the *Almagest* of Ptolemy.

9

Homer the Map-Maker

I and my predecessors . . . are right in
regarding Homer as the founder of the
science of geography.

Strabo, *Geography*, 1.1.10

Carrying little more than backpacks and with feet shod in
stout boots, adventurous travellers might yet again travel
the length and breadth of Greece along ancient ways with only
the *Iliad* and the stars as a guide. They would use the shapes of
constellations as their maps, and astronomy would determine
the points of the compass for them. They would be following in
the footsteps of men and women who trod the same paths three
thousand years ago, and probably even long before that time.

Little is known of the early history of cartography, but maps
inscribed on clay tablets and dating to *c.* 2300 BC have been
found in Babylonia. There is also evidence to support the view
that people of the Aegean travelled and traded beyond the
boundaries of Greece and Asia Minor long before Homer's
day. The presence of remains from Mycenae, in the
Peloponnese, in distant Cyprus and the discovery off south-
west Turkey of a wreck laden with cargo from afar, and dated
to 1316 BC (see Chapter 1), are some small testimony to a wide
trading network. Maps are vital today, and in one form or

another would have been no less important for distant travel in ancient days.

That Homer used his astronomical knowledge to draw skymaps[1] of the lands and islands of the Aegean is an idea in keeping with his skills in employing narrative to preserve learning about the skies. It is not proposed that he was the sole inventor of this geographical mapping system, but, unlike his fellow poet-scientists, he chose to preserve it, and his astronomy, in oral epic poetry. The magnitude of this achievement can be seen by looking at a modern star chart and asking the question, How on Earth – or in Heaven – could this multitude of stars and constellations ever be seen as maps? Nevertheless, such a link is exactly what Homer did manage to establish. Like much of his other scientific work, the initial idea is simple, but the composition and execution are rather more complex.

It has already been mentioned in Chapter 1 that Homer must have known something of the art of navigation, since in the *Odyssey* (5.271) Odysseus receives sailing instructions. Navigation was also a problem for Jason and the Argonauts, whose story predates Homeric epic. In the *Voyage of Argo*, Jason and his companions could not find their way home with the Golden Fleece. Jason, however, took the advice of his friend Argus:

'Think of a time when the wheeling constellations did not yet exist; when one would have looked in vain for the sacred Danaan race . . . Now we are told that from this country [Egypt] a certain king set out . . . and made his way through the whole of Europe and Asia, founding many cities as he went . . . to this day Aea stands, with people in it descended from the very men whom that king settled there. Moreover they have preserved tablets of stone which their ancestors engraved with maps giving the outlines of the land and sea and the routes in all directions.'

Argus finished and the goddess gave her blessing to the route he had proposed by sending them a sign. With cries of joy they saw ahead of them a trail of heavenly light, showing them the way to go.[2]

At first, Homer's combined role of astronomer and geographer may appear unusual, but it is one that was perpetuated by his scientific successors in later centuries, culminating in Ptolemy. The important bearings of north, south, east and west can quite easily be deduced from astronomical observations. During the night, the circumpolar stars wheel about the north celestial pole, whose position is indicated by the north pole star. By day, the highest point of the Sun's path across the sky, reached at noon, is due south. At the vernal and autumnal equinoxes the sun rises due east and sets due west. Chapter 1 illustrated how these bearings have been preserved by standing stones in Europe and alignments of temples and pyramids in Egypt.

Another link between astronomy and geography is that both stars in the sky and towns and other prominent features on Earth need to be charted. After Homer, both disciplines used an equatorial system of latitude and longitude to divide the globe of the Earth and the globe of the heavens. The 'latitude' of geography is known in astronomy as 'declination', while 'longitude' becomes 'right ascension'. Mathematics and geometry were the tools applied to solve common problems in both studies.

Among the early astronomer-geographers was Anaximander (c. 611–547 BC), who is said to have introduced the sundial to Greece, and who made an early attempt at a map of the inhabited world. Some two hundred years or so after Homer, Aristagoras of Miletus made another map of the world (c. 500 BC), and this was enlarged with a commentary by Hecataeus of Miletus (also c. 500 BC), who considered the Earth to be a flat disc. It was not until the fourth century BC that the astronomer Eudoxus of Cnidus claimed that the Earth was spherical. Hipparchus (born c. 190 BC) was possibly the greatest Greek astronomer of all, and is also recognized as the founder of a system that used mathematical principles to determine the position of places on the Earth's surface and specified their latitude and longitude.

The 'last great contribution to geographical science in antiquity'[3] was made by Ptolemy (c. AD 100–170), who, as with astronomy, had again become entwined with a subject familiar to Homer many centuries earlier. As well as the *Almagest*, his great astronomical treatise, Ptolemy created the *Guide to Geography*, an eight-volume work containing maps, the approximate latitudes and longitudes of eight thousand places, and instructions on how to prepare maps of the world.[4] Ptolemy's *Geography*, like his *Almagest*, had an influence on Western thought which was so lasting that in the fifteenth century AD Christopher Columbus wrongly estimated the distances to India and China because of errors made by Ptolemy.

Homer's reputation as a geographer survived him for several centuries, though he was not immune from criticism. The astronomer-geographer Eratosthenes (275–194 BC) repudiated the value of poets, including Homer, as geographers, likening poetry to a 'fable-prating old wife', whose purpose was to charm and not to instruct.[5] Eratosthenes, who assembled the geographical material of the famous library at Alexandria in Egypt into a scientific system and produced another map of the world, was in turn criticized by Hipparchus. Two centuries later, the noted geographer Strabo (c. 64 BC to AD 20) joined the argument over the role in science of the ancient poets and leaped to the defence of Homer:

> First, I say that both I and my predecessors, one of whom was Hipparchus himself, are right in regarding Homer as the founder of the science of geography.
>
> . . . as everyone knows, the successors of Homer in geography were also notable men and familiar with philosophy. Eratosthenes declares that the first two successors of Homer were Anaximander, a pupil and fellow-citizen of Thales, and Hecataeus of Miletus; that Anaximander was the first to publish a work on geography, a work believed to be his by reason of its similarity with other writings.
>
> Assuredly, however, there is need of encyclopaedic learning for the study of geography, as many men have already stated; and

Hipparchus, too, in his treatise *Against Eratosthenes*, correctly shows that it is impossible for any man, whether layman or scholar, to attain the requisite knowledge of geography without the determination of the heavenly bodies and of the eclipses which have been observed.

Geography . . . unites terrestrial and celestial phenomena as being very closely related, and in no sense separated from each other as Heaven is high above the Earth.[6]

The way in which Strabo defends Homer is curious. He devotes pages of argument in support of the poet, and draws together the strands of astronomy and geography as complementary studies. He refers to the importance of a knowledge of astronomy in the study of geography, and says that in the past myths had been used as a teaching medium. On the other hand Strabo does not record specific examples of Homer's astronomical and geographical learning, other than using the places named by the poet in the *Iliad* and the *Odyssey* as his guide. By the time of Strabo or soon afterwards, literature and science, which had been united under Homer and the poet-singers, must have split asunder, and so they remain today. Nevertheless, the quotations from Strabo show that Homer's scientific reputation was still alive in human memory as late as *c.* AD 20, when the geographer died.

The rediscovery of Homeric geography is believed to have been the first of Edna Leigh's many achievements in her research into the purpose of his epics, as we noted in Chapter 4. As early as the 1950s, she had noted that certain constellation shapes matched, to a greater or lesser degree, specific geographical areas of Greece and Asia Minor named in the *Iliad*. It would be a mistake, however, to consider Homeric geography in the light of modern techniques of highly detailed map-making. Homer uses constellation shapes to create mental images of the topography of different regions; used with features observed on the ground, these would have enabled travellers to find their way. Naturally, some of these shapes are a better 'fit' than others. Homer's maps are mainly concerned with showing the way through natural features such as ranges of hills, mountain passes and river valleys. Some of the maps are topographical and show details such as an island (Salamis) or a river valley (Sparta and the Eurotas valley); others are schematic, rather like the commonly used maps of the London Underground or the Paris Métro. As with modern road signs, the constellations do not give every twist and turn of the route between various points, but use straight lines.

Homeric geography may help to explain why many constellations bear no resemblance to their names. Legend has it that Cassiopeia takes the form of a queen sitting in a chair. However, the constellation more closely resembles the letter W than the shape of a woman, and is a workable schematic drawing of the roads of Euboea, the regimental home of the tribe of Abantes. By no stretch of the imagination does Auriga, the home of King Nestor, resemble a 'charioteer', but its outline identifies a pentagonal route system in his homeland of Messini in the southern Peloponnese.

The question arises of why the Homeric geographical tradition used the constellation shapes commonly believed to have been devised in Mesopotamia centuries and even thousands of years before. It would have been difficult beyond measure to import star patterns from a different culture and impose upon them the geographical features of Greece and Asia Minor. One could suggest, heretically, that these familiar shapes had their origins elsewhere, and were perhaps created by an ancient nomadic people who lived in Greece and Asia Minor. Those lands were not empty of human life before the rise of the Minoans and Mycenaeans, and people were farming there from the eighth millennium. This line of argument would go on to suggest that the constellations known to these very ancient people were subsequently adopted by the Mesopotamians.

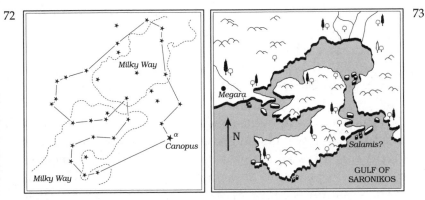

Fig. 72 *Argo Navis, home of Great Aias of the island of Salamis. The star Canopus not only represents Great Aias but may also represent the ancient capital of the island.*

Fig. 73 *The shape of the island of Salamis compares well with the shape of Argo Navis (fig. 72). The towns marked are given their modern names.*

In the Catalogue of Ships, Homer lists the homelands of the twenty-nine Greek and sixteen Trojan regiments who fought at Troy. Out of the forty-five constellations represented by the regiments, Homer uses at least thirty-six of them to create maps of Greece, Asia Minor, Crete and other Greek islands. A simple example of how Homeric geography works in practice can be seen below by comparing the ancient constellation of Argo Navis (fig. 72) with the island of Salamis (fig. 73), which the Catalogue of Ships says is the home of Great Aias and his men.

Until it was subdivided in the eighteenth century, Argo Navis was the largest constellation, and is thus an apt place in the sky for Great Aias, the 'largest' and second most powerful warrior on the battlefield of Troy. Its outline is similar to that of Salamis, and its brightest star, Canopus, is the second brightest in the heavens – a fitting personal star for Great Aias. Canopus would also be expected to mark the position of the island's ancient capital, believed also to have been called Salamis, which Strabo said 'faces towards Aegina and the south wind'.[7] The site has never been found, and it has been said that Strabo made an error. However, Canopus is in the same relative position in Argo Navis as the old city was said to be on Salamis, so Strabo may have been correct after all.

A small number of other constellations have been more difficult to associate with land areas, because, although Homer gives general areas of origin of regiments, he does not include enough information to draw meaningful maps. Examples of such areas are Ciconia, Halizoni and Mysia, which even today are relatively unknown. Of the Cicones, for instance, Homer says only that 'Euphemus, son of Troezenus, the son of Ceos, was captain of the Ciconian spearsmen' (2.846). This economical description lists neither towns nor ships, but a search of the *Iliad* for other clues and references to the Cicones has linked them astronomically to stars within Capricornus. Even so there is not enough information to construct a map. Three prominent constellations – Orion, Virgo and Libra – are not associated with land areas at all, but are important as the homes in the sky of the sons of King Priam (Orion), the sons of the Trojan elder Antenor (Virgo) and the sons of other Trojan elders (Libra).

Apart from the dozen or so skymaps to which Edna Leigh devoted the early stages of her study, there are other sketches of constellations and land areas in the now yellowing pages of her papers (see the end of this chapter for a full list of her identifications). These latter links appear to have been made in a burst of creative energy when ideas flowed so quickly there was no time for long explanations, only for cryptic notes relating mythology to geography. It is from this collection that the examples in the rest of this chapter have been taken. The style adopted for presentation is to use two diagrams: a star chart of a constellation, with prominent stars marking the sites of towns, and a geographical sketch of each area, using the ancient names.

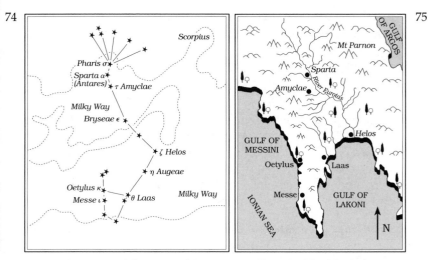

Fig. 74 Scorpius and the towns of the Eurotas valley. The 'shoulders' of Scorpius mark the headwaters of the valley.

Fig. 75 Sparta, home of Menelaus, and the Eurotas valley.

Scorpius and Lacedaemon

And those that dwelt in Lacedaemon, lying low among the hills, Pharis, Sparta, with Messe, the haunt of doves; Bryseae, Augeae, Amyclae, and Helos upon the sea; Laas, moreover, and Oetylus; these were led by Menelaus. (2.581)

The constellation Scorpius's representation of the kingdom of Menelaus is an excellent example of a topographical map following a route through natural features. The 'broad shoulders' of the constellation are the headwaters of the river Eurotas. Scorpius' brightest star in the 'chest' is Antares and represents Sparta (figs. 74 and 75). The valley of the Eurotas lies between the Parnon and Taygetos mountains, and the river forms the thin 'body' of the constellation. The spindly 'leg' and 'tail' of Scorpius make up the outline of the peninsula between the gulfs of Lakoni and Messini. The exact whereabouts of only four of the sites of Lacedaemon mentioned by Homer are known for certain: other locations were suggested by Edna Leigh.

Leo and Agamemnon's kingdom

Those who held the strong city of Mycenae, rich Corinth and [strong] Cleonae; Orneae, Araethyrea, and Sicyon, where Adrastus reigned of old; Hyperesia, high Gonoessa, and Pellene; Aegion and all the coast-land round and about Helice; these sent a hundred ships under the command of King Agamemnon, son of Atreus. (2.569)

Figs. 76 and 77 relate the towns and roads of Agamemnon's kingdom to the constellation Leo.

Mycenae, with its famous 'Lion Gate', lies just north of the Gulf of Argos, on a natural route that leads northward through a valley to Cleonae, near Nemea, and on to ancient Corinth and the Gulf of Corinth. Travelling along the coast westward from ancient Corinth, the route is hemmed in between the southern shores of the Gulf of Corinth and the steeply rising Peloponnese mountains before turning south. Travelling north-east from Corinth, a route passes through the once powerful town of Megara and Eleusis on the road towards Athens.

Modern astronomers who draw charts of Leo do not always agree about how stars should be connected to form an image of the constellation of the Lion. Stars in the 'head' can be connected to give the impression of an animal's head, but the same stars may also be presented as the Sickle of Leo, the name by which the asterism made up of α, η and γ (the handle) and ζ, μ, ϵ and λ (the blade) is commonly known. There is generally more agreement on the shape of the Lion's body, but at least one source gives a different view. Leo's legs also vary in number from four to none. The same stars are always there, of course – it is only how the human imagination joins them to form a memorable image that differs.

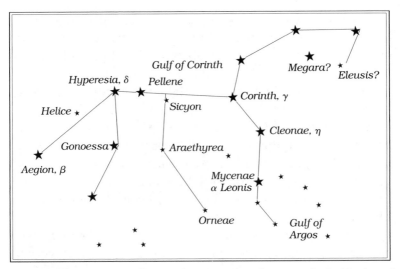

Fig. 76 The constellation of Leo, with stars representing towns in the kingdom of Agamemnon (fig. 77).

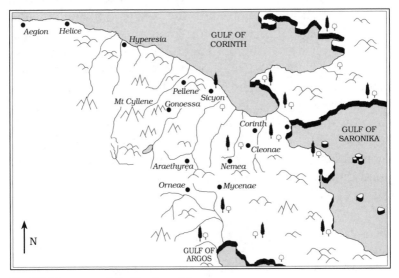

Fig. 77 Mycenae and the kingdom of Agamemnon.

Ptolemy gave Leo a head, legs and tail, and placed Regulus, α Leonis, the star that marks the city of Mycenae, at the 'heart' of the constellation. Edna Leigh created a skymap from this image of Leo.

The towns listed first in the Catalogue of Ships – Mycenae, Corinth and Cleonae – are marked by stars in Leo's chest. Homer uses the epithet 'strong' before Mycenae to indicate the brightness of Regulus, α Leonis, the brightest star on the ecliptic, and the epithet 'rich' before Corinth indicates the golden yellow star of Leonis close to 40 Leonis. Butler omits the usual epithet of 'strong' before Cleonae (η Leonis), another easily seen star.

There is a second route through the Peloponnese to the Gulf of Corinth, west of the one on which are sited Mycenae, Cleonae and Corinth. This goes through the middle of the 'body' of Leo, and on it lie the towns of Orneae, Araethyrea and Sicyon. Along the 'back' of Leo, westward along the Gulf of Corinth from Sicyon, are the towns of Pellene and Hyperesia, and the coastal tract continues down the 'tail' through Helice and Aegion. The site of 'high Gonoessa' is not known, but Gonoessa is proposed as a town marked by a star in the rear leg of Leo; there is a natural route through the mountains following a river in that region, and the epithet of 'high' suggests that Gonoessa was in an elevated place.

Edna Leigh believed that stars in the head of Leo projected Agamemnon's interests beyond his immediate kingdom, and point the road to the towns of Megara and Eleusis which then led on to Athens.

Ursa Major and Troy

Ursa Major and its asterism of seven stars that make the Big Dipper form the best-known constellation in the heavens, which on each clear night of the year is seen circling around the pole star. After many centuries of rebuilding on top of older

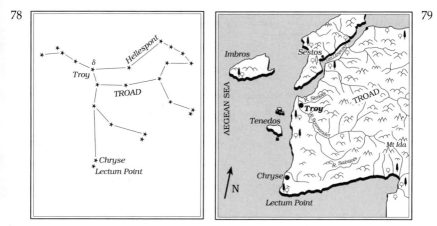

Fig. 78 *Ursa Major, constellation of the Troad.*

Fig. 79 *Troy and the Troad.*

ruins, Troy stood out prominently on the plain, and it has been described as looking like an inverted pudding basin. That same image was evoked by Homer when he wrote, 'Now there is a high mound before the city, rising by itself upon the plain. Men call it Batieia [Thorn Hill], but the gods know that it is the tomb of lithe Myrine. Here the Trojans and their allies divided their forces' (2.811). The steep mound of Myrine led to the identification of Troy with Ursa Major and the seven stars of the Dipper or Plough, the 'mound' being the four bright stars in the bowl of the Dipper, and the three stars in the handle representing a channel which led down to the sea. In the third millennium, Troy was a sea-girt headland, but by 1250 BC, as the estuaries of the rivers Simois and Scamander became silted up, the city was landlocked and about a mile from the sea. Archaeologists have found evidence of a canal or channel leading from the city to the Hellespont. When the Dipper is placed on a map of that part of Turkey it defines not only Troy, but the coastline from the Hellespont down to Lectum Point (figs. 78 and 79).

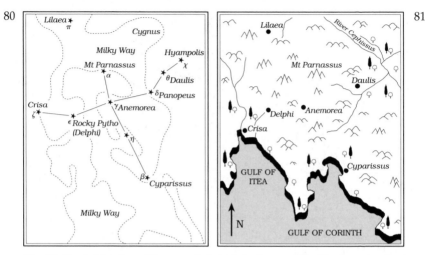

Fig. 80 *The bright stars of Cygnus that mark the towns of the Phocians stand out against the background of the Milky Way.*

Fig. 81 *The area of rocky Pythos (Delphi) and Mount Parnassus.*

Cygnus and the Men from Phocis

The Phocians were led by Schedius and Epistrophus . . . These were they that held Cyparissus, rocky Pytho, holy Crisa, Daulis, and Panopeus; they also that dwelt in Anemorea and Hyampolis, and about the waters of the river Cephissus, and Lilaea by the springs of the Cephissus. (2.517)

Cygnus lies in a beautiful part of the sky in which scores of stars shine against the background of the Milky Way, and it represents the region around holy Delphi (rocky Pytho) in Phocis (figs. 80 and 81). Deneb, α Cygni, nineteenth among the twenty brightest stars in the sky, is identified with the 8,065-foot peak of Mount Parnassus. Of this region, Achilles says, 'My life is more to me than all the wealth of Ilius . . . or all the treasure that lies on the stone floor of Apollo's temple beneath the cliffs of [rocky] Pytho' (9.405). Delphi, or Delfoi, was certainly the site of a treasury of wealth on Earth and in the skies.

Using a telescope, the astronomer Herschel counted 331,000 stars in the region of ϵ Cygni,[8] and even naked-eye observers have claimed to identify almost 200 stars set against the back-cloth of the Milky Way. 'Rocky', describing Pytho, is a code-word Homer uses elsewhere to denote parts of the sky with an abundance of stars related to mountainous land areas.

Auriga and the Kingdom of King Nestor

The men of Pylos and Arene, and Thryon where is the ford of the river Alpheus; strong Aepy, Cyparisseis, and Amphigeneia; Pteleus, Helos, and Dorion . . . These were commanded by Nestor, knight of Gerene. (2.591)

Homer refers to Nestor's home as being in 'sandy Pylos', giving an impression of long golden beaches – an impression that is supported by a tour along the coast in that part of his kingdom. But 'sand', like 'wheat' and 'barley', is a Homeric codeword for the Milky Way, whose bright band of light runs through Auriga.

The pentagonal shape of Auriga reflects the ancient routes along the coast and through the mountains of Messini (figs. 82 and 83). It is not known where the town of Thryon stood, except that Homer says it was at a ford on the river Alpheus. The shape of Auriga suggests it stood somewhere near Olympia.

Although the present-day location of Dorion is known, it will be noticed that Dorion is missing from the skymap in fig. 82. This is because there is no star in the outline of Auriga to mark it. However, Homer says that at Dorion the Muses had caused Thamyris the bard to lose his power of song (2.594), which possibly signifies the absence of a star.

Aquarius and the Hellespont, with the Lands of Asius

They that dwelt about Percote and Practius, with Sestos, Abydus, and Arisbe – these were led by Asius, son of Hyrtacus, a brave

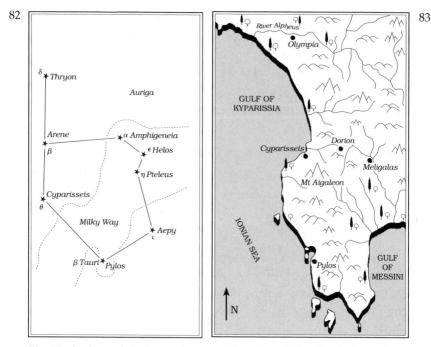

Fig. 82 *Auriga and the known towns of King Nestor. The site of Thryon is not known, but Homer says it is on the Alpheus river, perhaps near Olympia.*

Fig. 83 *Pylos and Messini. The shape of Auriga indicates ancient routes along the coast and through the mountains.*

> commander . . . whom his powerful dark bay steeds, of the breed that comes from the river Selleis, had brought from Arisbe'. (2.835)

The popular image of Aquarius is that of a young man carrying a water jar, but one wonders whether there is a touch of Greek humour in allocating the constellation to a 'water carrier' of greater dimensions. To try to realistically match the shape of the constellation to a human figure is difficult, but when that shape is compared to that of the Hellespont it makes much more sense (figs. 84 and 85). In a manner of speaking, the Hellespont too is a 'water carrier', as it takes the fast-flowing waters of the Black

84 85

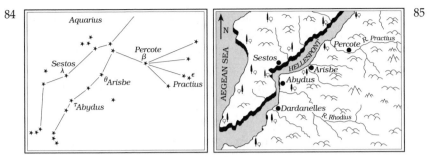

Fig. 84 *Aquarius and towns on the shores of the Hellespont.*

Fig. 85 *The Hellespont.*

Sea into the Aegean. Stars of Aquarius identify towns and a river system on the shores of the Hellespont, or the Dardanelles as it is more commonly known today.

Draco/Ursa Minor and Boeotia

These [Boeotians] dwelt in Hyria and rocky Aulis, and . . . held Schoenus, Scolus, and the highlands of Eteonus, with Thespeia, Graia, and the fair city of Mycalessus. They also held Harma, Eilesium, and Erythrae; and they had Eleon, Hyle, and Peteon; Ocalea and the strong fortress of Medeon; Copae, Eutresis, and Thisbe the haunt of doves; Coronea, and the pastures of Haliartus; Plataea and Glisas; the fortress of Thebes the less; holy Onchestus with its famous grove of Poseidon; Arne rich in vineyards; Midea, sacred Nisa, and Anthedon upon the sea. (2.494)

The previous examples of skymaps have been relatively uncomplicated, because in each case only a small number of towns had to be placed. This brief sampling, however, would not be complete unless the Boeotians, the most complex regiment of all, were included.

The Boeotians were the first Greek force listed in the Catalogue of Ships, but, although numerous in terms of commanders (5), places (30), ships (150), and men (1,750), their regiment does not play a pivotal role in the fighting. Later commentators have suggested the Boeotians had the honour of leading the Greek forces on to the beaches of Troy because a later Boeotian poet changed the running order of the catalogue. Edna Leigh said the answer to this anomaly lies in astronomy, for stars in Draco and Ursa Minor play very important roles in Homer's geography as well as in his exposition of the precession of the equinoxes. The tail of Draco includes Thuban, the pole star of the *Iliad*, together with other stars which have indicated the north celestial pole in the past and will again in the future. The star α Ursae Minoris, known as Polaris, is the present pole star.

For Homeric geography, our interest lies in the thirty Boeotian towns listed in the catalogue, and the sites of some twenty-six of them are believed to be known. Those which remain unlocated are Eteonus, Eilesium, Arne and Nisa – although, as we explain later, 'sacred Nisa' can be equated with Mount Helicon. Of the sites found, more than half are close to the modern counterparts of Aulis, Hyria, Thebes, Plataea, Eutresis, Thespeia, Coronea, Haliartus, Onchestus, Copae, Midea, Thisbe, Anthedon, Glisas, Harma and Mycalessus. Over the centuries there has been considerable investigation into the location of the places listed by Homer, but much relies on ancient Greek sources: the geography of Strabo and that of Pausanias (AD 143–176). The rediscovery of Homeric astronomy may eventually lead to the identification of the presently unknown sites.

The assocation between Draco and the populous region of Boeotia was the result of a painstaking piece of research by Edna Leigh, and as will be seen shortly there is a mythological link between the region and the legend of a dragon. Geographically, stars in the tail of Draco indicate a route from Aulis, via Hyria to the south-west towards Thebes the less and Plataea (figs. 86 and 87). The body sweeps northward to Eutresis, Thespeia and Coronea, and then twists southward

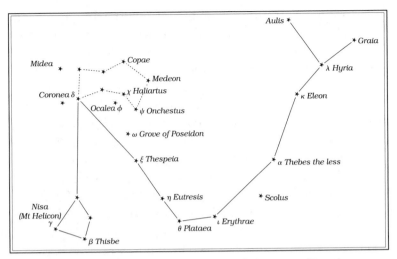

Fig. 86 The stars of Draco associated with the towns of Boeotia.

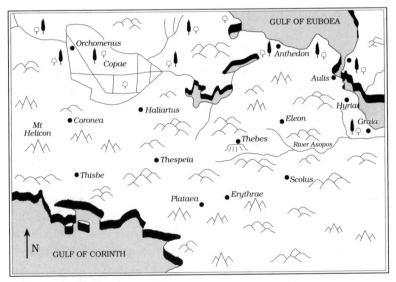

Fig. 87 Boeotia and its known Homeric sites.

257

towards the two brightest stars of Draco in the dragon's head. Although the location of 'sacred Nisa' is unknown, γ Draconis is the brightest star of Draco, and could associate Mount Helicon – sacred to Apollo and the favourite haunt of the Muses – with Nisa. If Mount Helicon represents sacred Nisa, the entry for the Boeotians in the Catalogue of Ships concludes with a full sweep of Boeotia from Nisa (Mount Helicon) in the south-west to Anthedon by the sea in the north-east. In fig. 86 an elliptical pattern of stars conveniently marks out the now largely drained lake of Copae, with stars representing towns on its shores. On the map of Boeotia and its towns, fig. 87, an indication of the size of the lake is provided by the drainage ditches shown.

The stars of Ursa Minor, which may not have been seen as a separate constellation at the time of Homer,[9] form the smaller topographical area of Anthedon, Glisas, Harma and Mycalessus (figs. 88 and 89).

THE GROVE OF POSEIDON

The precession of the equinoxes and its subsequent effects are never far from the surface of *Homer's Secret Iliad*. One place listed among the Boeotian dwelling-places is the grove of Poseidon, which has never been given a positive identification on land, and perhaps never will be. However, according to the sequence of the stars of Draco, the grove of Poseidon would be represented by ω Draconis. Although relatively faint, with a magnitude of 4.8, its location in the skies is of the upmost importance for the understanding of the universe. It is the nearest star to the centre of the northern precessional circle, or the pole of the ecliptic as it is known in astronomical terms, around which the heavens appear to circle (see Chapter 7). By invoking the name of Poseidon, an observation of fundamental importance for astronomical theory is established in unforgettable terms. The site of the grove of Poseidon, if it ever existed, would have lain well inland, and perhaps shows

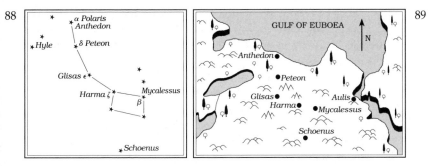

Fig. 88 Stars of Ursa Minor and the towns they represent.

Fig. 89 Towns linked to the stars of Ursa Minor.

another aspect of the god of the seas and his astronomical significance.

DRACO'S HEAD AND THE CADMEA

Edna Leigh associated the head of Draco with the mythology of the legendary foundation of the city of Thebes by Cadmus, and went on to expand her work to establish the association of the head and the long and sinuous body and tail of Draco with a skymap for Boeotia. She wrote of the legend of Cadmus, who is accredited with bringing writing to Greece:

> The Phoenician Cadmus consulted the Delphic oracle, who directed him to build a town wherever a certain cow might lie down. This Cadmus did, but only after he had battled with a nervous dragon. The perspicacious beast was apparently sceptical of allowing strangers to imbibe information from a spring in the area and thereby become able to learn sacred lore. Cadmus and the dragon came to terms, however, and he built his city, overlying an earlier town, sometime after 2000 BC. The natural terrain determines the shape of the Cadmea, the acropolis in the citadel of Thebes, which lies on a hill some 150 feet high. To the west is the deep ravine of a small river, the Dirce; just east of the Cadmea is another stream, the Strophia. These flow northward and converge

at a place near the walls which once surrounded the whole of the city. The Cadmea fits very neatly into the head of Draco, who coils himself about the fortified area of the king's palace and doubly guards its treasures.

It is no accident that Cadmus' cow chose to lie down at Thebes, nor is it coincidental that Cadmus' acropolis can be pictured as surrounded by Draco. In an old legend, Cadmus once killed a dragon and planted his teeth, and from the soil there sprang armed warriors who fought among themselves until only five remained. Known as the Sown Men, the five afterwards helped Cadmus build and defend his city.

Edna Leigh also associated those five men with Arcesilaus and the four other Boeotian leaders.

Regions Associated with Constellations

Following is a list of the regions of Greece and Asia Minor and the constellations to which they are linked. They are listed in the order of the roll-call of the regiments in the Catalogue of Ships (2.494).

GREEK REGIMENTS

The twenty-nine Greek contingents can be subdivided into five geographical areas, corresponding to groups of constellations in the same region of the sky. However, there are exceptions, and the reason for this is not currently known.

Central Mainland Greece

Draco and Ursa Minor = Boeotia. Part of Draco = Aspledon. Cygnus = Phocis. Cepheus = Locris. Cassiopeia = Euboea. Southern Cross (later Camelopardalis) = Athens. Argo Navis = Salamis. All of these constellations are circumpolar, with the important proviso that they did not all go around the same pole. Draco, Ursa Minor, Cygnus, Cepheus and Cassiopeia circle around the North Pole, but Argo Navis and the Southern

Cross long ago went around the South Pole. The reasons for the Southern Cross and Argo Navis being included in this group are complex. Edna Leigh's explanation of this apparent contradiction is in her commentary on Book 10, and involves precession and the death of Dolon (Camelopardalis).

The Peloponnese
Perseus = Argos and Tiryns. Leo = Mycenae. Scorpius = Lacedaemon. Auriga = Pylos. ι Leonis = Arcadia. Cancer = Buprasion and Elis. All in this group are either fully in the zodiac or have stars bordering the zodiac. For example, Perseus' feet extend to the zodiac, and El Nath, a star in the zodiac, was in ancient times in Auriga. ι Leonis, a bright star in the rear leg of the Lion, is within the boundaries of the zodiac, and on Earth Arcadia is thought to have been part of Agamemnon's empire, and could have been included in the constellation.

Western Mainland Greece and the Cephallonian Isles
Canes Venetici = Dulichium and the Echinean Islands. Boötes = Cephallonia. Aries = Aetolia. Although the three land areas associated with these contingents are geographically near to one another, Aries is far distant from Boötes and Canes Venetici in the skies. Aries, however, leads into the next group.

Islands of the Aegean
Taurus = Crete.   Pisces = Rhodes.   Triangulum = Syme. Andromeda = Calydnian Islands. All of these constellations are in the same area of the sky and represent islands of the Aegean seas.

Eastern Mainland Greece
Canis Major = Pelasgian Argos. Ursa Minor = Phylace and Pyrasus.   Pegasus = Lake   Boebe.   Hercules = Methone,

Thaumacia, Meliboea and Olizon. This group of four appears to have a south–north and east–west configuration. Ophiuchus = Tricce, Ithome and Oechalia. Lyra = Ormenion, Hypereia, Asterion and Titanus. Aquila = Argissa, Gyrtone, Orthe, Helone, Oloösson. Sagitta = Enienes and Peraebians. Delphinus = Magnetes. These last two are found within or close to the Milky Way.

THE TROJANS

With only a few exceptions – such as the strongholds of Troy, Adresteia, Percote, Paphlagonia, Maeonia and Caria – Homer gives less information about places associated with the Trojan regiments than he does about the Greek homes. A number of Trojan allies came from remote parts of Asia Minor, and lack of information about these regions makes them difficult to associate with constellations. Orion, Virgo and Libra are not associated with any geographical region.

Trojan Regiments
Ursa Major = Troy and the Troad. Sagittarius = Zelia. Equuleus = Adresteia. Aquarius = Percote. Ophiuchus = Pelasgian Larissa. Capricornus = Thrace. ϵ Aquarii = Ciconia. Centaurus and Crux = Paeonia. Lupus = Paphlagonia. Crater = Halizoni. η and μ Geminorum = Mysia. Corvus = Phrygia and Ascania. Piscis Austrinus = Maeonia. α and β Centauri = Caria. Gemini = Lycia.

Epilogue

> When the mode of learning changed in the
> following ages, and science was delivered in
> a plainer manner, it then became as reason-
> able in the more modern poets to lay it aside,
> as it was in Homer to make use of it.
>
> Alexander Pope in the Preface to
> *The Iliad of Homer* (1715–20)

We have seen that, as late as the first years of the Christian era, the geographer Strabo vigorously defended Homer's scientific reputation. This suggests that there was in the centuries after Homer at least a folk memory of his achievements as a scientist. We can speculate on two main reasons for Homer's eclipse in this area. The first was the spread of writing at about the time that he lived, which would immediately have much reduced the need for aids to memory and rendered the teaching roles of poet-singers redundant. Such, though, is the power and beauty of his epics that they lived on as popular entertainment and great literature long after their roles as allegories of the heavens had been forgotten. The quotation by Alexander Pope above suggests that he had a similar conception of the poet's former role.

The nature of science also changed after Homer, and his learning about the heavens, based on observations with the naked eye, may not have been considered accurate enough for the new generation of philosophers. No longer was it sufficient to observe the skies: the disciplines of mathematics and geometry were being used to try to explain the harmony of the heavens. In this new atmosphere of intellectual exploration, Homer's learning may have been considered homespun lore from the distant past.

We hope that our ideas about Homeric astronomy will lead not only to a revision of the history of astronomy, but also to an extension of the study of the purpose of mythology in Europe and elsewhere. We also hope this work will stimulate interest in classical and historical studies, and that lay people and professionals will share some of the pleasure and excitement that we have enjoyed.

Homer's Secret Iliad may at first arouse antagonism from those who have studied and read the epic as literature. But they should reflect that, far from undermining Homer's reputation, this book adds a whole new dimension to his genius. Homer remains supreme as a poet, and his status as the greatest single influence on Western literature is not diminished one jot by the idea of the epics being composed to preserve scientific knowledge. The advent of Homeric astronomy enhances his reputation still further, and he must now also be accorded a place of high rank among the great pioneers of science.

Homer's Secret Iliad is a distillation of the papers of Edna Leigh and of a lengthy astronomical commentary and a catalogue of 650 stars and constellations so far identified in the *Iliad*. It is self-contained, and covers the principal areas of Homeric astronomy. It is our belief that most if not quite every incident in the *Iliad* has an astronomical bearing, and narrative not included in *Homer's Secret Iliad* is examined in the commentary. This additional material has no startling new revelations, but expands upon those recorded here. Even so, there is narrative that has not yet succumbed to interpretation, but we are confident that it will do so in the future when those with a wider knowledge of naked-eye astronomy begin to examine the *Iliad*. Edna Leigh believed it would be fifty years before all of Homer's learning was understood. Meanwhile, work is continuing on compiling and editing her exciting research into the *Odyssey* and her excursions into the purpose of other myths.

As for Homer, he continues to be the person about whom nothing is known and much is speculated, but at least one romantic legend can be decently buried. Homer has been depicted as a blind storyteller, but no sightless man could have had such a vast knowledge of the heavens.

Appendix 1

The Greek Alphabet Used in Bayer's Star Classification

1. α alpha
2. β beta
3. γ gamma
4. δ delta
5. ε epsilon
6. ζ zeta
7. η eta
8. θ theta
9. ι iota
10. κ kappa
11. λ lambda
12. μ mu
13. ν nu
14. ξ xi
15. o omicron
16. π pi
17. ρ rho
18. σ sigma
19. τ tau
20. υ upsilon
21. φ phi
22. χ chi
23. ψ psi
24. ω omega

Appendix 2

Optical Aids and the Greeks

The detailed knowledge of the heavens acquired by the peoples who lived in pre-Homeric Crete and the lands and islands of the Aegean Sea is so remarkable that it is natural to raise the question of whether they had any form of viewing aid. In a short essay, Edna Leigh put forward an idea derived from the legend of the renowned Greek hero Perseus. Not only does he himself have a constellation in the sky, but others close by are those of his wife, Andromeda, her mother, Cassiopeia, and his winged horse, Pegasus. Edna wrote:

Among the myths which have come down to us from a remote period is the highly popular account of the Greek hero Perseus, a man who succeeded in cutting off the head of the fabulous Medusa, a girl with snakes for hair. Before he was born, Perseus' mother spent many years in a specially constructed, round bronze room which admitted only light and air. Later, both he and she were cast adrift in a closed box on the seas.

Perseus' adventures began when he undertook to bring Medusa's head to the king. Medusa was horrific, and her gaze turned to stone those who dared look at her – a hazard which Perseus avoided with the help of the gods. He looked not at Medusa but at her reflection in a highly polished round bronze shield, a special gift from Athene, the goddess of wisdom. For a time Perseus also held in his hand a unique and portable eye, one shared in common by three divinities. To carry out his conquest, Perseus also acquired an adamantine sickle, a pair of winged sandals, a cloak of invisibility and a magic case to hold Medusa's head. According to some writers, Perseus performed the beheading feat at night.

If, for the time being, we disregard the narrative and concentrate on some of the objects, we find among them the following: a special room for letting in light, an eye that could be held in the hand, a diamond-bright sickle, a mirror-like shield, a case and a snaky head. Let the room be one used for observing light; let the unusual eye be a lens, the bright shield a mirror, and the sickle a clear-cut image rather than an ugly blurred object, Medusa. The magic case might just possibly be a container holding light, a mirror, an image and a lens. Medusa might be a star, and the container a telescope.

So far as we know, Sir Isaac Newton was the first to use any form of reflecting telescope. When Newton made his, in A D 1688, he used a special mirror made of two parts copper and one part tin; ancient bronze was ordinarily nine parts copper and one part tin.

Apart from Perseus receiving his bright shield from Athene, we know nothing about it. Because Athene was patron of the arts, however, the shield she gave Perseus undoubtedly had some significance. What I have just outlined in no way proves that selected threads of the Perseus legend describe a telescope, but I do suggest they may. My purpose . . . is to indicate that a mythical story of fairy-tale quality may, in an unexpected manner, illustrate a few key words appertaining to astronomy.

Attached to her typewritten account of Perseus is a cutting of a letter written by Richard Hughes, of the United University Club, London, and published in the *New Scientist* on 5 May 1966. It includes the following paragraph about the Hyperboreans, a mythical race of people who worshipped Apollo and were said to live in the far north where the Sun rose and set only once a year:

> There is much else of scientific interest in this Hecataeus (b. circa 550 B C) fragment about the Hyperboreans, but a single example must suffice. We are told that there 'the Moon appears to be so close to the Earth that the mountains on it are clearly visible'. That seems to imply some sort of primitive telescope – for without one, how could anyone even have guessed there were mountains on the Moon to see? Now, a writer on the pre-Christian origins of the Grail legend has found that this prehistoric cult-object had alternative forms such as metal bowl and crystal ball, and both forms were used to kindle sacrificial fire (presumably by concentrating the rays of the sun). Both forms used in conjunction would indeed provide the elements of a Newtonian telescope . . . Pindar, moreover, says it was in the land of the Hyperboreans that Perseus slew the Gorgon with the aid of some such sacred curved mirror . . .

That two people thought along similar lines is perhaps only a curiosity at this point, but it might stimulate an investigation into other myths and legends that offer similar evidence for such devices.

A Short Glossary of Astronomical Terms

asterism   A distinctive group of stars within a constellation, such as the Plough or Dipper in Ursa Major.

celestial poles   The imaginary points at which the Earth's axis of rotation, if extended, would meet the celestial sphere. The star nearest to the north celestial pole at any one time is known as the pole star, and is a relatively stable point around which the heavens appear to rotate.

celestial sphere   An imaginary sphere of infinite size which appears to rotate around the Earth and on which the stars seem to be fixed. This geocentric view of the Earth at the centre of the universe was known to Homer and survived until overturned in A D 1543 by Copernicus' heliocentric theory, which placed the Sun at the centre of our solar system.

circumpolar constellations   Constellations that circle around the poles of the northern and southern hemispheres and do not set below the horizon.

conjunction   The apparent coming together in the sky of two or more planets. This is an effect caused by observing the alignment of planets separated by vast distances of space.

constellations   Astronomers today divide the stars into eighty-eight configurations such as Taurus, Virgo, Leo, Perseus, Hercules and Ursa Major. Homer divided the heavens into forty-five constellations.

cosmology   A theoretical or mythical account of the nature or origin of the universe.

eclipse   Lunar eclipses occur when the moon passes through the shadow of the Earth. Solar eclipses are really occultations (see below), and happen when the Moon is directly between the Earth and the Sun. Lunar and solar eclipses can be total or partial.

ecliptic   The apparent path of the Sun across the heavens over the course of a year.

equinox   At the vernal (spring) and autumnal equinoxes the sun is at the equator and there are equal hours of day and night.

fireball   An exceptionally bright meteor.

magnitude   The apparent brightness of stars is compared on a scale of magnitude – the higher the number, the fainter the star. The brightest star in the sky is Sirius, α Canis Majoris, with a magnitude of -1.4. Stars of up to about magnitude 6 can be seen with the naked eye.

meridian   An imaginary circle which passes around the celestial sphere through the celestial poles at the zenith of a given place.

meteors   Particles of matter, usually small, which burn up and produce a streak of light on entering the Earth's atmosphere.

meteor showers   A swarm of meteors that enters the Earth's atmosphere from a particular part of the sky. Regular displays can be seen at the same time each year, and are associated with the constellation from which they radiate: the Geminids, the Leonids, the Lyrids, the Orionids etc.

Milky Way   The galaxy of which our solar system is a part appears as a broad band of light which stretches across the celestial sphere.

nadir   The point on the celestial sphere directly below an observer at any point on Earth.

occultation   The phenomenon whereby a body nearer the Earth, such as the Moon or a planet, temporarily obscures a body further away, such as a star, as it passes across the sky.

precession of the equinoxes   The apparent slow westward shift of the two points where the ecliptic crosses the celestial equator, caused by the gravitational pull of the Sun and Moon on the Earth. As the equinox precesses, the constellations in which the Sun rises at the vernal and autumnal equinoxes change. Currently the Sun rises in Pisces at the vernal equinox, but in a few centuries it will rise in Aquarius.

radiant   The point on the celestial sphere from which meteors and meteor showers appear to radiate.

retrograde motion   The phenomenon whereby a planet can appear to stop and then move backwards along its path for a period of weeks before proceeding on its normal course.

solstices   At the summer solstice (longest day) in June, the Sun is at its maximum declination (highest point) in the sky of approximately 23½°. It is at its lowest at the winter solstice (longest night) in December.

supernova   A star that explodes with a cosmic outburst of such enormity that, for a time, it can be a prominent sight in the sky.

variable star   A star that fluctuates in brilliance or magnitude.

zenith   The point on the celestial sphere directly above the observer's head at any place on Earth.

zodiac   A band of sky sixteen degrees wide that contains the paths of the Sun, the Moon and all the planets. Homer's zodiac – like ours – was divided into twelve parts, each represented by a constellation in which the sun rose for one month each year.

Notes and References

Chapter 1: Astronomy and the Ancients

1. J. McKim Malville, Fred Wendorf, Ali A. Mazar and Romauld Schild, 'Megaliths and Neolithic astronomy in southern Egypt', *Nature*, vol. 392, 2 April 1998, pp. 488ff.
2. Comment reported by the Press Association, 1 April 1998.
3. *The Times*, 28 October 1996.
4. Richard Hinckley Allen, *Star Names, Their Lore and Meaning* (New York: Dover, 1963; first published as *Star Names and Their Meaning*, 1899), pp. 362, 109, 382.
5. Ibid., pp. 205, 353, 307, 77.
6. E. C. Krupp, *In Search of Ancient Astronomies* (London: Chatto & Windus, 1973), p. 186.
7. Anthony Aveni, *Stairways to the Stars: Skywatching in Three Great Ancient Cultures* (Chichester: John Wiley, 1997), p. 110.
8. 'American Indian pottery. South America', *Encyclopaedia Britannica*, 1998.
9. Gerald S. Hawkins, with John B. White, *Stonehenge Decoded* (London: Souvenir Press, 1965).
10. John North, *Stonehenge: Neolithic Man and the Cosmos* (London: HarperCollins, 1996), p. xxxvi.
11. Hesiod, *Theogony, Works and Days* (with Theognis, *Elegies*), trans. Dorothea Wender (Harmondsworth: Penguin, 1973), p. 71 (ll. 383–4).
12. Göran Henriksson and Mary Blomberg, 'Evidence for Minoan astronomical observations from the peak sanctuaries on

Petsophas and Traostalos', *Opuscula Atheniensia*, vol. XXI, no. 6 (1996), and Mary Blomberg and Göran Henriksson, 'Archaeoastronomical light on the priestly role of the king in Crete', *Acta Universitatis Upsaliensis*, 1993.

13. Michael Ovenden, 'The origin of the constellations', *Philosophical Journal*, vol. 3, no. 1 (1965), pp. 1–18.

14. Archie E. Roy, 'The origin of the constellations', *Vistas in Astronomy*, vol. 27 (1984), pp. 171–97.

Chapter 2: Preserving Learning in Epic

1. Peter Green, *A Concise History of Ancient Greece to the Close of the Classical Era* (London: Thames and Hudson, 1973), p. 10.

2. John Chadwick, *The Decipherment of Linear B* (Cambridge: Cambridge University Press, 1970), p. 132.

3. Ibid., p. 124.

4. W. F. Jackson Knight, *Many-Minded Homer*, ed. John Christie (London: Allen & Unwin, 1968), pp. 43–4.

5. See Milman Parry, *The Making of Homeric Verse: The Collected Papers of Milman Parry*, ed. Adam Parry (Oxford: Clarendon Press, 1970).

6. Review by Allison Pearson of George Psychoundakis, *The Cretan Runner* (London: John Murray, 1978; Penguin, 1998), *Daily Telegraph*, 13 June 1998.

7. Frances A. Yates, *The Art of Memory* (London: Routledge & Kegan Paul, 1966), p. 29.

8. Alexander Pope, preface to his translation (1715–20) of *The Iliad of Homer* (London: Grant Richards, 1902), p. xi.

9. Knight, *Many-Minded Homer*, pp. 124–5.

Chapter 3: Trojan War on Earth and in the Skies

1. Thomas W. Allen, *The Homeric Catalogue of Ships* (Oxford: Clarendon Press, 1921), p. 168.

2. Malcolm M. Willcock, *A Companion to the Iliad: Based on the Translation by Richmond Lattimore* (Chicago and London: University of Chicago Press, 1976), p. 23.

3. Gabrielle Camille Flammarion and André Danjon, *The Flammarion Book of Astronomy*, trans. Annabel and Bernard Pagel (New York: Simon & Schuster, 1964), p. 13.

Chapter 4: Warriors as Stars

1. Quoted in T. W. Allen, *The Homeric Catalogue of Ships*, p. 2.
2. T. Wynne Griffon, *Star Maps: Your Guide to the Night Sky* (London: Bison, 1992), p. 43.
3. Christian Heinrich Friedrich Peters and Edward Ball Knobel in *Ptolemy's Catalogue of Stars*, a revision of the *Almagest* (Washington: Carnegie Institute, 1915), p. 27.
4. Ptolemy's star and constellation catalogue (AD 125) is based on that of Hipparchus (b. 190 BC), of which no copy remains.
5. G. J. Toomer, ed., *Ptolemy's Almagest* (London: Duckworth, 1984), p. 15.
6. Both Homer and Ptolemy viewed Gemini as 'matchstick' twins. Homer in the *Iliad* saw them as the warriors Sarpedon and Glaucus, while Ptolemy saw them as the 'rear twin' and the 'advance twin'.
7. The ability of observers in Homer's day to discern small differences in magnitude was examined at length in Edna Leigh's notes.
8. *The Iliad*, trans. E. V. Rieu (Harmondsworth: Penguin, 1950), p. vii.
9. Willcock, *A Companion to the Iliad*, pp. 57–8.
10. Jean-Pierre Verdet, *The Sky: Order and Chaos* (London: Thames and Hudson, 1992), p. 34.
11. R. H. Allen, *Star Names*, p. 255.

Chapter 5: Warriors as Constellations

1. See also R. H. Allen, *Star Names*, p. 433.
2. Giorgio de Santillana and Hertha von Dechend, *Hamlet's Mill: An Essay Investigating the Origins of Human Knowledge and its Transmission through Myth* (Boston: Godine, 1983), p. 384.
3. J. B. Sidgwick, *Introducing Astronomy* (London: Faber and Faber, 1961), p. 200.
4. Ibid., p. 152.

Chapter 6: Gods in the Heavens

1. Hesiod, *Theogony*, p. 27 (ll. 124–30).
2. David Sacks, *Encyclopaedia of the Ancient Greek World*, ed. consultant Oswyn Murray (London: Constable, 1995), p. 204.
3. *Sky and Telescope*, December 1993, 'Newsnotes', pp. 13–14, reporting on work by Kevin D. Pang and John A. Bangert.

4. Translation by Marjorie M. J. Rigby.
5. For different reasons, Homer uses the phrase 'white-armed' as an epithet for Helen of Troy – see page 107.
6. Photographs of the Moon taken through a telescope can appear confusing, as the image is reversed and upside down. As a guide to the seas, the almost circular patch on the top right of the Moon when viewed by the naked eye is Mare Crisium.
7. Georg Autenrieth, *A Homeric Dictionary*, trans. from the German by Robert P. Keep (1876), rev. Isaac Flagg (Norman: University of Oklahoma Press, 1958).

Chapter 7: The Changing Heavens and the Fall of Troy

1. Pliny, *Natural History*, trans. H. Rackham (Loeb Classical Library; Cambridge, Mass.: Harvard University Press, 1958), II.95.
2. R. H. Allen, *Star Names*, p. 1.
3. Willcock, *A Companion to the Iliad*, pp. 104–5.
4. Harald A. J. Reiche, 'The language of archaic astronomy', in Kenneth Brecher and Michael Feirtag, *The Astronomy of the Ancients* (Cambridge, Mass.: MIT Press, 1980), pp. 155, 159.
5. M. I. Finley, *The World of Odysseus* (Harmondsworth: Penguin, 1967), p. 172.
6. Data from Joachim Schultz, *Movement and Rhythms of the Stars* (Edinburgh: Floris Books, 1986 (first published in Germany in 1963)), p. 218.

Chapter 8: Homer's Earth-Centred Universe

1. Translation by Marjorie M. J. Rigby.
2. Flammarion and Danjon, *The Flammarion Book of Astronomy*, p. 12.
3. John North, *The Fontana History of Astronomy and Cosmology* (London: Fontana, 1994), p. 67.

Chapter 9: Homer the Map-Maker

1. Edna Leigh used the word 'cartogram' for her own maps associating constellations with land areas; 'skymap' is a word devised to define drawings based on notes and thumbnail sketches in her papers.

2. Apollonius of Rhodes, *The Voyage of Argo: The Argonautica*, trans. E. V. Rieu (Harmondsworth: Penguin, 1959), p. 154 (IV.260).
3. 'Geographia', in Harry Thurston Peck, ed., *Harper's Dictionary of Classical Literature and Antiquities* (New York: Cooper Square Publishers, 1965 edn), p. 724.
4. 'Mapping and surveying', *Encyclopaedia Britannica*, 1994–7.
5. Strabo, *The Geography of Strabo*, trans. Horace Leonard Jones (Loeb Classical Library; Cambridge, Mass.: Harvard University Press, 1917–32), 1.2.3.
6. Ibid., 1.1.10–12.
7. Ibid., 9.1.9.
8. Sidgwick, *Introducing Astronomy*, p. 237.
9. R. H. Allen, *Star Names*, p. 448.

Sources

The prime source for original material in this book is the papers of Edna F. Leigh.

Bibliography and Further Reading

THE EPICS OF HOMER

The nineteenth-century translation of the *Iliad* by Samuel Butler (London: Longman, 1898; available on-line at http://classics.mit.edu/Homer/iliad.html) has been used for quotations in this book. Butler preferred to use Latin names for some of Homer's characters, e.g. Ulysses for Odysseus, and Venus for Aphrodite etc. These have been changed to their more familiar forms, and his capitalization has been harmonized with that used elsewhere. The line numbers given for the start of passages referred to are likely to vary slightly from translation to translation.

Other translations of the *Iliad* and the *Odyssey* used for reference are:

The Iliad, trans. Robert Fagles (London: Penguin, 1991)

The Iliad of Homer, trans. Andrew Lang, Walter Leaf and Ernest Myers (1882; reprinted London: Macmillan, 1961)

The Iliad of Homer, trans. Richmond Lattimore (Chicago and London: University of Chicago Press, 1961)

The Iliad, trans. A. T. Murray, with Greek Text (Loeb Classical Library; Cambridge, Mass.: Harvard University Press, 1924–5, reprinted 1960)

The Iliad of Homer, by Alexander Pope (1715–20; reprinted London: Grant Richards, 1902)

The Iliad, trans. E. V. Rieu (Harmondsworth: Penguin, 1950)
The Odyssey of Homer, trans. S. H. Butcher and A. Lang (1870; reprinted London: Macmillan, 1963)
The Odyssey, trans. E. V. Rieu (Harmondsworth: Penguin, 1946)

GEOGRAPHY

Allen, Thomas W., *The Homeric Catalogue of Ships* (Oxford: Clarendon Press, 1921)
Dicks, D. R., ed., *The Geographical Fragments of Hipparchus* (London: Athlone Press, 1960)
Simpson, R. Hope, and Lazenby, J. F., *The Catalogue of Ships in Homer's Iliad* (Oxford: Clarendon Press, 1970)
Strabo, *The Geography of Strabo,* trans. Horace Leonard Jones (Loeb Classical Library; Cambridge, Mass.: Harvard University Press, 1917–32)
Wood, Michael, *In Search of the Trojan War* (London: BBC Books, 1985)

MYTHOLOGY AND REFERENCE

Autenrieth, Georg, *A Homeric Dictionary,* trans. from the German by Robert P. Keep (1876), rev. Isaac Flagg (Norman: University of Oklahoma Press, 1958)
Cary, M., and Nock, A. D., eds., *The Oxford Classical Dictionary* (Oxford: Clarendon Press, 1949)
Chadwick, John, *The Decipherment of Linear B* (Cambridge: Cambridge University Press, 1970)
Graves, Robert, *Greek Myths* (Harmondsworth: Penguin, 2 vols., 1960)
Green, Peter, *A Concise History of Ancient Greece to the Close of the Classical Era* (London: Thames and Hudson, 1973)
Kirk, G. S., *Nature of Greek Myths* (Harmondsworth: Penguin, 1974)
Knight, W. F. Jackson, *Many-Minded Homer,* ed. John Christie (London: Allen & Unwin, 1968)
Hesiod, *Theogony, Works and Days* (with Theognis, *Elegies*), trans. Dorothea Wender (Harmondsworth: Penguin, 1973)
Hornblower, Simon, and Spawforth, Antony, eds., *The Oxford Classical Dictionary* (Oxford: Oxford University Press, 3rd edn, 1996)
Peck, Harry Thurston, ed., *Harper's Dictionary of Classical Literature and Antiquities* (New York: Cooper Square Publishers, 1965 edn)

Sacks, David, *Encyclopaedia of the Ancient Greek World*, ed. consultant Oswyn Murray (London: Constable, 1995)

Willis, Roy, ed., *World Mythology* (New York: Henry Holt, 1996)

Willcock, Malcolm M., *A Companion to the Iliad: Based on the Translation by Richmond Lattimore* (Chicago and London: University of Chicago Press, 1976)

Yates, Frances A., *The Art of Memory* (London: Routledge & Kegan Paul, 1966)

ASTRONOMY

Bone, Neil, *Observer's Handbook of Meteors* (London: George Philip, 1993)

Couper, Heather, and Henbest, Nigel, *The Stars* (London: Pan, 1988)

De Callatay's Atlas and the Skies, trans. Sir Henry Spencer Jones (London: Macmillan, 1959)

Flammarion, Gabrielle Camille, and Danjon, André, *The Flammarion Book of Astronomy*, trans. Annabel and Bernard Pagel (New York: Simon & Schuster, 1964)

Griffon, T. Wynne, *Star Maps: Your Guide to the Night Sky* (London: Bison, 1992)

Moore, Patrick, *Guinness Book of Astronomy* (London: Guinness Publishing, 1992)

Rey, H. A., *The Stars* (Boston: Houghton Mifflin, 1962)

Ridpath, Ian, ed., *Norton's 2000* (Harlow: Longman Scientific and Technical, 1989)

Schultz, Joachim, *Movement and Rhythms of the Stars* (Edinburgh: Floris Books, 1986 (first published in Germany in 1963))

Sidgwick, J. B., *Introducing Astronomy* (London: Faber and Faber, 1961)

Tancock, E. O., ed., *Philip's Chart of the Stars* (London: George Philip, 1990)

Verdet, Jean-Pierre, *The Sky: Order and Chaos* (London: Thames and Hudson, 1992)

HISTORY OF ASTRONOMY

Allen, Richard Hinckley, *Star Names, Their Lore and Meaning* (New York: Dover, 1963; first published as *Star Names and Their Meaning*, 1899)

Aveni, Anthony, *Stairways to the Stars: Skywatching in Three Great Ancient Cultures* (Chichester: John Wiley, 1997)

Barrow, John D., *The Artful Universe* (London: Penguin, 1997)

Brecher, Kenneth, and Feirtag, Michael, *Astronomy of the Ancients* (Cambridge, Mass.: MIT Press, 1980)

Dicks, D. R., *Early Greek Astronomy to Aristotle* (London: Thames and Hudson, 1970)

Hoskin, Michael, ed., *Cambridge Illustrated History of Astronomy* (Cambridge: Cambridge University Press, 1997)

Krupp, E. C., *In Search of Ancient Astronomies* (London: Chatto & Windus, 1973)

North, John, *The Fontana History of Astronomy and Cosmology* (London: Fontana, 1994)

——, *Stonehenge: Neolithic Man and the Cosmos* (London: HarperCollins, 1996)

Pannekoek, A., *History of Astronomy* (London: Allen & Unwin, 1961)

Santillana, Giorgio de, and Dechend, Hertha von, *Hamlet's Mill: An Essay Investigating the Origins of Human Knowledge and its Transmission through Myth* (Boston: Godine, 1983)

Thurston, Hugh, *Early Astronomy* (New York: Springer-Verlag, 1994)

Toomer, G. J., ed., *Ptolemy's Almagest* (London: Duckworth, 1984)

Walker, Christoper, ed., *Astronomy before the Telescope* (London: British Museum Press, 1996)

Computer programs used to show the heavens at the time of Homer and long before included *SkyGlobe*, a shareware program by KlassM Software (1990), *The Sky* (1992) by Software Bisque, and *Red Shift*. Diagrams in the book show the skies from the latitude of Athens in the correct historical periods.

Index

Achaeans, 51

Achernar (star), 185–6

Achilles: withdraws from battle and sulks, 50, 69, 192–3, 219; as character in *Iliad*, 51, 53; kills Hector, 51, 72–4, 76–7, 118–19, 129, 177, 179–80; as son of Thetis, 52, 150, 153, 185; identified with star Sirius, 60, 65–6, 69, 73–4, 83, 118–20, 125, 147, 149–53, 154, 191–5, 200–1, 218; staffs and spears, 67, 214; and passage of stars around Earth, 70, 72, 231–2; identified with constellation Canis Major, 72, 75–6, 84, 108, 125, 128, 135, 147–53, 176, 180, 198, 201, 231–2; Priam observes, 73, 190; return to battle, 77, 191, 193–4, 207, 219, 220; killings and woundings, 90, 100, 184; helmet, 104; kills Andromache's father and brothers, 108–9; horses, 113–14; 'hut', 117, 149, 195, 197; death, 118; epithets, 119; Deeds of Glory, 124; arming of, 128, 150, 152, 165, 181, 192–3, 198, 207; and death of Patroclus, 135; efforts to persuade to return to battle, 149, 192, 196; troubled sleep, 150, 224, 231; as warrior, 150–3; wounded, 151–2; forces Trojans to flee, 164; shield, 165, 181, 198–206; protected by Athene, 176; and creation of new constellation, 196–7; receives gifts, 196; and Leo constellation, 250; on Phocis, 252

Adresteia, 155, 262

Adrestus, 94, 100

Aegean Sea, 239–40, 255, 261

Aegion, 250

Aeneas: and Meriones, 45, 173; as character in *Iliad*, 52; identified with Spica and Virgo, 71, 88, 120, 128, 155, 179, 221, 224, 228; and Diomedes, 144, 224, 228; protected by mother (Aphrodite), 177, 179, 228; kills Medon, 216

Aesculapius, 110

Aethra (Helen's handmaid), 107

Aetolia, 154, 261

Agamemnon, King of Mycenae: leads army against Troy, 50; as character in *Iliad*, 51, 53; supposed mask, 54; and inevitability of fall of Troy, 67–8; staffs and spears, 67, 172, 181, 214; and Achilles' sulk, 69, 192; home in Mycenae, 82, 248–50; identified with Regulus and Leo, 83–4, 86, 88, 107, 109, 115, 120–1,

281

occultations, 59, 163–4, 176, 179,
197, 237–8
Oceanus (deity), 164, 236
Oceanus (river), 145, 199, 206
octaëteris, 38–9
Odius, 97, 99, 155
Odysseus: raids Trojan lines, 28,
233; as character in *Iliad*, 51;
staffs and spears, 67, 141;
armour and shield, 84, 97, 102,
125, 143; identified with
Arcturus and Boötes, 85–6, 88,
102, 120, 124–5, 128–9, 137–43,
154, 174, 176, 194, 233; personal
star, 86; wounded, 98, 141–2;
killings and woundings, 102–3;
physique, 126–7; qualities, 137;
accused of cowardice, 138–9;
attempts to persuade Achilles to
return to battle, 149, 196;
protected by Athene, 176;
suggests delay before resuming
battle, 193
Odyssey: on funeral customs, 5;
Edna Leigh's interpretation of,
10, 265; astronomical content, 15;
sailing instructions and
navigation in, 37, 240;
memorized, 46–7; original texts,
49; place names in, 243
Oechalia, 262
Olizon, 262
Oloösson, 262
Olympus, Mount: as home of Zeus,
63, 71, 234–5; as zenith, 235–6
omens and signs, 31–2, 187–9
Ophiuchus (constellation), 98,
110–11, 262
oral tradition: and memorizing
techniques, 46–7, 219–21, 263
Orion (constellation): shape, 20;
bow, 21, 110; Hector identified
with, 60, 72–7, 92, 105–6, 108,
121, 128–9, 142, 155, 171, 224,
229–31, 246; Paris identified
with, 71–2, 89, 92, 109–11, 121,

128–32, 177–8, 225, 226, 246;
Priam's sons associated with,
91–2; in Homer and Ptolemy, 98;
boundaries, 116–17, 194; meteor
shower, 122; and omen, 188; and
Achilles' shield, 200; passage,
223–4, 230, 233; not associated
with geographical region, 262
Ormenion, 262
Orneae, 250
Orthe, 262
Ossa, Mount, 79
Ouranos, 156
Ovenden, Michael, 39–40

Paeonia, 262
Paiawon (Paian), 44
Pandarus: identified with
Sagittarius, 71, 97, 113–14, 155,
209–10, 217–18, 220, 224, 227–9;
wounds Menelaus, 90, 99, 103,
125, 128, 130, 209–10, 217, 220,
221, 227; wounds Diomedes, 98,
146, 224, 228; killed, 210, 227–9
Pang, Kevin D., 162
Panthous, 91–2
Paphlagonia, 262
Paris (son of Priam): birth and
prophecy on, 49–50; judges most
beautiful goddess, 50, 52, 161,
166, 174; as character in *Iliad*, 52,
53; identified with Betelgeuse and
Orion, 71–2, 84, 89, 92, 109, 120–1,
124–5, 128–32, 177–8, 225–6,
230–1; cloak, 84; woundings and
killings, 98, 99, 110, 112–13, 133,
146, 216; escapes, 121; duels with
Menelaus, 129–31, 224–7, 234;
challenges Greek warriors, 130,
224; protected by Aphrodite, 177,
225; Helen joins in bed, 223,
225–6; follows brother Hector,
224, 229–31
Parnassus, Mount, 252
Parry, Milman, 46
parselene (mock moons), 169

Trojan Horse, 51
Tros, 152
Troy: Blegen excavates at, 12; siege
and capture of, 50–2, 219;
inevitability of fall, 52, 67–8, 207,
214, 220; site (Hissarlik) and
status, 52–5, 86; burning (c. 1220
BC), 53; identified with Ursa
Major, 65, 67, 73, 92–4, 111, 115,
180, 207, 213–14, 236, 250–2, 262;
regiments, 262

Ursa Major (constellation): and fall
of Troy, 65–6, 67–8, 207, 214;
identified with Troy, 65, 67, 73,
92–4, 111, 115, 180, 207, 213–14,
236, 250–2, 262; in Homer and
Ptolemy, 98; as constellation,
123–4, 196; as funeral wagon,
123; in Achilles' shield, 200
Ursa Minor (constellation), 255–61

Valestinon (formerly Pherae), 79, 80
Vega (star), 110, 111, 120, 215–16
Ventris, Michael, 6, 44
Venus (planet): identified with
Aphrodite, 62–3, 158, 161–3, 165,
177–8; visibility, 63; passage, 70
Virgo (constellation):
identifications, 71, 92, 111, 128,
224, 246; associated with
Antenor's sons, 91, 92; and
summer solstice, 220–1; setting,
224; not associated with
geographical region, 262
Volos (formerly Iolcus), 79, 80

Willcock, Malcolm M., 196

women: in Troy, 106–8
Wood, Florence (née Reid),
daughter of Edna Leigh, 3, 12,
14, 90
Wood, Kenneth, 8
writing, Sumerian, 30

Xanthus (horse), 113
Xanthus (Milky Way), 164–5, 184

Yates, Frances, 46–7
year: measurement and division
of, 31–3, 39

zenith, 223, 234–5
Zeus (deity): in Linear B texts, 44;
promise to Thetis of Achilles'
glory, 52; status as deity, 62–3;
non-intervention in death of son
Sarpedon, 67, 211; intervenes in
Iliad action, 71, 164; and
Erechthonius, 131; fed by she-
goat, 133; and heavenly bodies
and movements, 157–60, 162,
234; rebels against Cronus, 157;
relations with Hera, 166–9, 178;
gives birth to Athene, 174;
threatens Athene and Hera, 175;
and Athene's defiance, 176;
instructs Hermes, 180; throws
down Hephaestus, 181; and
omens, 188–9; admonishes gods,
224, 234–5
zodiac: defined, 22–4; and
precession, 65; Mercury in, 180;
divisions, 195; as memory aid,
219–22; and planetary
movements, 238

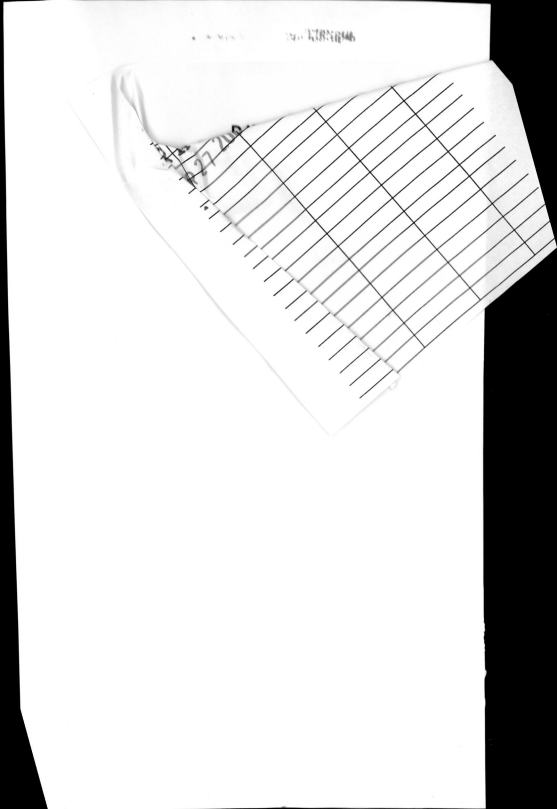